Praise for *Good Questions for Math Teaching, High School . . .*

Good Questions for Math Teaching is a masterpiece that will leave teachers wondering how they ever wrote lesson plans without it. The resource not only provides invaluable tips and questions that can be effortlessly integrated into every lesson but it also demonstrates how good questions can serve as a catalyst for driving deeper learning of mathematical relationships.

—*Samuel Rhodes, Assistant Professor of Elementary Mathematics Education, Georgia Southern University, Statesboro, Georgia*

Good Questions for Math Teaching offers dozens of rich tasks, posed as questions, about a range of high school content, from Number and Quantity to Calculus. *Good Questions* works well as "do-now" tasks, unit summary tasks, or for general formative assessment. The authors really know how to support students to dig deeper into content that might otherwise seem procedural and how to make thinking about a variety of mathematical content interesting. These rich tasks have the potential to shift students away from being passive receivers of information toward being engaged mathematical thinkers. My one critique is that the book should be called *Great* (not just *Good*) *Questions!*

—*Michelle Cirillo, teacher educator and researcher, Department of Mathematical Sciences, University of Delaware, Newark, Delaware*

The tasks in *Good Questions for Math Teaching* are sorted by unit and are open-ended to support teachers in moving beyond problems where the only "answer" is $x = 5$ while simultaneously encouraging students to consider a variety of paths to a possible solution. We all want to build experiences for students that invite them into the joy that is learning about and playing with math, and this resource helps do just that; it's an amazing collection of open-ended problems.

— *Heather Sugrue, Academic Dean and Upper School Mathematics Teacher, Milton Academy, Milton, Massachusetts*

Good Questions for Math Teaching delivers what it promises—good questions that get students (and you!) thinking more deeply about math. The questions also get students talking about mathematics and talking with math—using mathematical terms, representations, and diagrams to explain their thinking. Some questions in this book are great warm-up problems that can replace procedural "do-now" exercises with thought-provoking questions to investigate. Other questions can be used as the basis for a full lesson-long investigation. All of these questions are thought provoking, allow students choice in solution method, and focus on important mathematical concepts.

—*William Zahner, Interim Director, Center for Research in Mathematics and Science Education, San Diego State University, San Diego, California*

Good Questions for Math Teaching provides everything teachers need to incorporate effective questioning into their own lessons—fantastic sets of questions, guidance on how to use them with students, and a window into the mathematical richness that underlies the high school curriculum.

—*Ziv Feldman, coauthor of* Talk Moves: A Teacher's Guide for Using Classroom Discussions in Math, Middle and High School

HIGH SCHOOL

GOOD QUESTIONS FOR MATH TEACHING

WHY ASK THEM AND WHAT TO ASK

NANCY ANDERSON · LESLIE DIETIKER · GREGG REILLY

Series Editor, Nancy Anderson

Math Solutions
Boston, Massachusetts, USA

Math Solutions, a division of Houghton Mifflin Harcourt
www.mathsolutions.com

Math Solutions is a division of Houghton Mifflin Harcourt.

Cataloging-in-Publication Data on file with Library of Congress.

ISBN-13: 978-1-935099-78-9

Executive Editor: Jamie Ann Cross
Production Manager: Denise A. Botelho
Editorial Assistant: Kirby Sandmeyer
Interior Design and Composition: Publishers' Design and Production Services, Inc.
Photo Credits: Nancy Anderson (Portrait Simple); Leslie Dietiker (Boston University); Gregg Reilly (Gregory White)
Cover Design: Susan Olinsky Design
Question mark art created by sdecoret / Shutterstock; Shopping Cart icon on back cover created by Freepik from www.flaticon.com

Printed in the United States of America.

1 2 3 4 5 6 7 8 9 10 0304 27 26 25 24 23 22 21 20
4510007031 ABCDE

A Message from Math Solutions

We at Math Solutions believe that teaching math well calls for increasing our understanding of the math we teach, seeking deeper insights into how students learn mathematics, and refining our lessons to best promote students' learning.

Math Solutions shares classroom-tested lessons and teaching expertise from our faculty of professional learning consultants as well as from other respected math educators. Our publications are part of the nationwide effort we've made since 1984 that now includes:

- more than five hundred face-to-face professional learning programs each year for teachers and administrators in districts across the country;
- professional learning books that span all math topics taught in kindergarten through high school;
- videos for teachers and for parents that show math lessons taught in actual classrooms;
- on-site visits to schools to help refine teaching strategies and assess student learning; and
- free online support, including grade-level lessons, book reviews, inservice information, and district feedback, all in our Math Solutions Online Newsletter.

For information about all of the products and services we have available, please visit our website at *www.mathsolutions.com.* You can also contact us to discuss math professional development needs by calling (877) 234-7323 or by sending an email to *info@mathsolutions.com.*

We're always eager for your feedback and interested in learning about your particular needs. We look forward to hearing from you.

Math Solutions.
FOUNDED BY MARILYN BURNS
From Houghton Mifflin Harcourt.

Brief Contents

(Continued)

How to Access Online Resources

To access the downloadable reproducibles referenced in the text, please see (Roman numeral) page xx for directions, visit mathsolutions.com/myonlineresources, and register for an account (even if you have one with our bookstore). Use key code **GQHS** to register this product and download.

Contents

(Continued)

How to Access Online Resources

To access the downloadable reproducibles, please see (Roman numeral) page xx for directions, visit mathsolutions.com/myonlineresources, and register for an account (even if you have one with our bookstore). Use key code **GQHS** to register this product and download.

Reproducibles

The following reproducibles are referenced throughout the text. These reproducibles are also available in a downloadable, printable format. For access, visit www.mathsolutions.com/myonlineresources and register your product using the key code **GQHS**. See (Roman numeral) page xx for more detailed instructions.

Reproducible 1	Equations Venn Diagram
Reproducible 2	Area Diagram A
Reproducible 3	Area Diagram B
Reproducible 4	Graph of Expressions *A* and *B*
Reproducible 5	Points (x, y) Venn Diagram
Reproducible 6	Annie's Graph
Reproducible 7	Graph of Transformed Functions
Reproducible 8	Area of Rectangle
Reproducible 9	Expressions Venn Diagram
Reproducible 10	Solutions Venn Diagram
Reproducible 11	Gerald's Graph
Reproducible 12	Inequalities Venn Diagram
Reproducible 13	AB's Graph

(Continued)

(Continued)

(Continued)

(Continued)

Acknowledgments

Peter Sullivan and Pat Lilburn's *Good Questions for Math Teaching: Why Ask Them and What to Ask, Grades K–5, Second Edition*, as well as Lainie Schuster and Nancy Anderson's *Good Questions for Math Teaching: Why Ask Them and What to Ask, Grades 5–8, Second Edition*, have been the model for much of our work in this book. Both books have been extremely popular with teachers because of their content and format and compatibility with a variety of curricular approaches. We hope this book, focused on algebra through introductory calculus, will be just as helpful to secondary math teachers and their students.

We teach in an era of wonderful resources and standards-based curricula. Our teaching has continued to develop and improve because of them. We would like to acknowledge the following series, publications, and their authors, for providing us with the inspiration for many questions in this book: NCTM's Developing Essential Understanding series, Pearson's CME Project textbook series, and CPM Educational Program's Core Connections textbook series.

We would like to give special thanks to our reader John Banderob, whose edits and suggestions were always timely, specific, and encouraging. We also wish to thank our Editor, Jamie Cross; our Editorial Assistant, Kirby Sandmeyer; and our Production Manager, Denise Botelho for their patience, "gentle nudges," and skill in guiding this project from its earliest stages all the way to its final production. Thanks to our careful copyeditor Anne Sauvé, and to Ethan Botelho and Mathematical Expressions for their detailed and mathematically keen eyes!

Finally, we wish to thank our students and colleagues for inspiring and delighting us with your hard work, courage, and insights. This book is written for you.

—Nancy Anderson, Leslie Dietiker, and Gregg Reilly

How to Use This Resource

Why This Resource?

Our goals of education are for our students to think, to learn, to analyze, to criticize, and to be able to solve unfamiliar problems, and it follows that good questions should be part of the instructional repertoire of all teachers of mathematics. As we work to emphasize problem solving, the application of concepts and procedures, and the development of a variety of thinking skills in our mathematics curricula, it becomes vital that we pay increased attention to the improvement of our questioning techniques in mathematics lessons. As teachers of mathematics, we want our students not only to understand what they think but also to be able to articulate how they arrived at those understandings. Developing productive questions can help focus learning on the process of thinking while attending to the study of content (Dantonio and Beisenherz 2001, 60). Good questions created and posed by teachers ultimately become powerful tools for student learning.

In *Good Questions for Math Teaching, High School* we describe the features of good questions, show how to create good questions, give some practical ideas for using them in your classroom, and provide many good questions that you can use in your mathematics program. By asking careful, purposeful questions, teachers create dynamic learning environments, help students make sense of math, and unravel common incorrect responses.

How Is This Resource Organized?

This resource is divided into two parts: Part I, "The Importance of Good Questions," and Part II, "Good Questions to Use in Math Lessons."

Part I

In Part I we explain what we mean by "good questions," walk you through different methods to create your own good questions, and provide necessary support for how to use good questions in your classroom.

Part II

In Part II we offer examples of good questions, broken down by topic, for you to select from and use in your classroom. Each question is accompanied by notes to support your teaching.

When Should I Use These Questions?

Here are a few of our favorite scenarios for when to use good questions.

> **When to Use Good Questions**
>
> - Use good questions to supplement your mathematics curriculum.
> - Use good questions for warm-up routines or for review.
> - Incorporate good questions into homework and/or assessments.

Use Good Questions to Supplement Your Mathematics Curriculum

The questions in this book are designed as a supplement to your mathematics curriculum. Good questions can be used as the basis for an entire lesson, either as a lesson that stands alone or as a part of a unit of study. As you progress through a unit of study, you may want to ask your students questions in this book that correspond to that unit. Embedding questions from this book in your lessons may further enhance student learning and understanding. When studying rational exponents, for example, you may wish to refer to the questions in this book to support, extend, or enrich your present curriculum. If current practice has your students solving word problems that have only one answer, you may wish to ask students questions that allow for multiple answers—and approaches—such as Chapter 4, Question 7 on page 27 (*Replace the letters* a *and* b *with a positive value for* a *and a negative value for* b *in the expression* a^b *to create two different expressions such that one simplifies to a positive, rational number and the other to a positive, irrational number*). By constructing an answer to this question, students gain valuable experience with rational exponents and also gain insights into the differences between negative exponents and negative numbers.

Use Good Questions for Warm-Up Routines or for Review

Another way to use this book is to use the questions as a daily warm-up activity or "do-now" activity for the start of math class. You may choose a question that corresponds to the current unit or a question that requires students to review a particularly challenging or important skill or concept. Students can talk about the question with their peers before you convene the class with a discussion of students' answers. This routine makes good use of transition time while immediately focusing math class on reasoning and communication.

Incorporate Good Questions into Homework and/or Assessments

In addition to using these questions during instructional time, you could assign them for homework or incorporate them into your assessments. Typically, math homework and assessment practices tend to focus on skills and/or closed questions that require recall of what was learned in class. If your students usually complete a review worksheet of the day's lesson, attach (or even substitute) a question from this book that will require them to think beyond what they have learned, to connect an understanding from a previous lesson, or to confront a misconception that may have arisen during class. In addition, if students' assessments normally include computations and problems for which there is only one correct answer, adapt those assessments to include some questions from this book that will allow students to think creatively about the mathematics they are learning.

How to Access Online Resources

1. Go to mathsolutions.com/myonlineresources and log in if you already have an account. If you do not, click or tap the Create New Account button at the bottom of the Log In form.
2. Create an account, even if you have created one with the Math Solutions bookstore. You will receive a confirmation email when your account has been created.
3. Once your account has been created, you will be taken to the Product Registration page. Click Register on the product you would like to access (in this case, *Good Questions for Math Teaching, High School*).

[Key Code GQHS]

4. Enter key code **GQHS** and click or tap the Submit Key Code button.
5. Click or tap the Complete Registration button.
6. To access the reproducibles at any time, visit your account page.

More Resources in This Series!

We are excited to share this resource, *Good Questions for Math Teaching, High School* as one of several resources in the Good Questions series. For more titles, see www.mathsolutions.com.

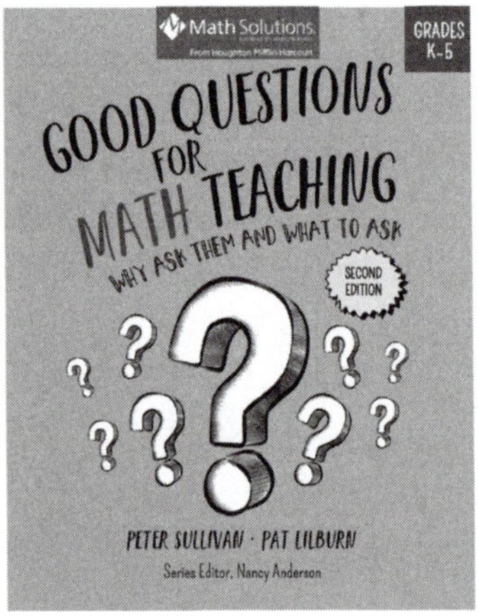

Grades K–5
ISBN: 978-1-935099-76-5

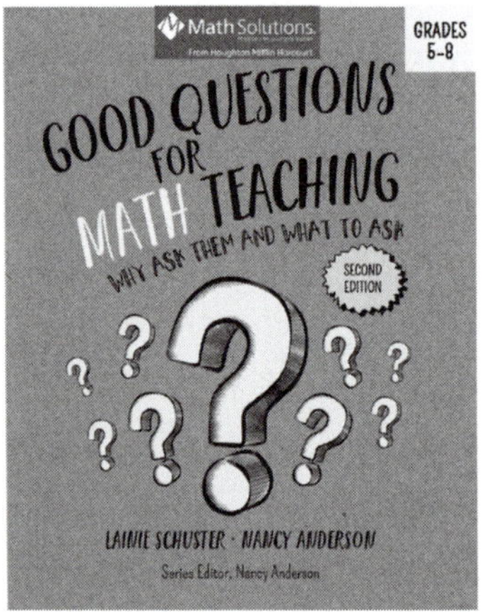

Grades 5–8
ISBN: 978-1-935099-77-2

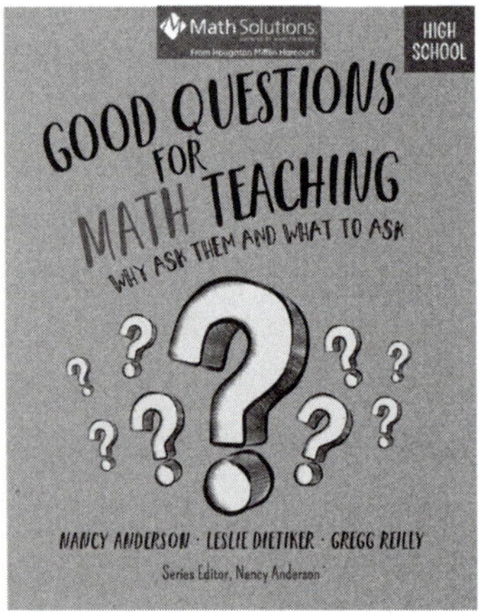

High School
ISBN: 978-1-935099-78-9

PART I

The Importance of Good Questions

In Part I, we explain what we mean by good questions, walk you through different methods to create your own good questions, and provide necessary support for how to use good questions in your classroom. Remember, the power of questioning is ultimately in the answering. As teachers, we not only need to ask good questions to get good answers but need to ask good questions to promote the thinking required to give good answers.

CHAPTER 1

What Are Good Questions?

During the course of a normal school day teachers ask many questions. In fact, something like 60 percent of the things said by teachers are questions and most of these are not planned. One way of categorizing questions is to describe them as either open or closed. Closed questions are those that simply require an answer or a response to be given from memory, such as a description of a situation or object or the reproduction of a skill. Open questions are those that require a student to think more deeply and to give a response that involves more than recalling a fact or reproducing a skill.

Teachers are usually skilled at asking open questions in content areas such as language arts or social studies. For example, teachers often ask students to interpret situations or justify opinions. Unfortunately, however, in mathematics lessons, closed questions are much more common.

Questions that encourage students to do more than recall known facts have the potential to stimulate thinking and reasoning. To emphasize problem solving, application, and the development of a variety of thinking skills, it is vital that we pay more attention to improving our questioning in mathematics lessons. Teachers should use questions that develop their students' higher levels of thinking.

Good Questions for Math Teaching, High School looks in more detail at a particular type of open question that we call a "good" question. Good questions can set the stage for meaningful classroom discussion and learning. When we ask good questions in math class, we invite our students to think, to understand, and to share a mathematical journey with their classmates and teachers alike. Students are no longer passive receivers of information when asked questions that challenge their understandings and convictions about mathematics. They become active and engaged in the construction of their own mathematical understanding and knowledge.

Careful, intentional, and mindful questioning is one of the most powerful tools a skillful teacher possesses (Costa and Kallick 2000,

34). So, what do good questions "look" like? Let's consider four main features.

Four Main Features of Good Questions

1. Good questions help students make sense of the mathematics.
2. Students can learn by answering good questions, and the teacher learns about each student from the attempt.
3. Good questions may have several acceptable answers.
4. Good questions are accessible to all students.

Feature 1: Good Questions Help Students Make Sense of the Mathematics

Too often mathematics is presented in a closed or rote manner such that students are given a procedure (or told about a concept) and then asked to absorb that information without much reflection or exploration. Good questions, on the other hand, require more than remembering a fact or reproducing a skill. Their open-endedness requires the application of facts and procedures while simultaneously encouraging students to make connections and generalizations. Good questions help students see mathematics as a subject that is based on sense making. In turn, good questions empower students to think of themselves as capable sense makers. Let's look at a couple of scenarios to help further explore this.

Scenario 1: Quadratic Equation Questions

Think about the following two questions—which one do you feel is more powerful in helping students make sense of mathematics? Why?

> **Question A:** *Find the number of real solutions to the equation* $0 = (x - 2)^2 + 3$.
>
> **Question B:** *Tuhina looked at a quadratic equation in vertex form and thought to herself, "Based on the graph of the related function, I know the equation has no real solutions." What might be the equation Tuhina was looking at?*

Question A might be considered a typical kind of question about quadratic equations that is asked of students in the study of algebra. These questions typically present students with a quadratic equation and ask them to find its number of real solutions. While we would not want these questions to disappear entirely, we also recognize that these questions have their limitations. Specifically, they do not empower students to make decisions based on what makes sense or consider connections between important mathematical concepts.

What if we took Question A and opened it up? That's where Question B—what we consider a "good question"—comes into play. Question B invites students to use the characteristics of the graph of a quadratic relationship to reason about the solutions to a related equation while also making decisions about how to organize their thinking, eliminate possibilities, and find more than one possible answer. When students are given many opportunities to derive their own strategies, try them out, make adjustments, and justify their reasoning, they gain knowledge, thinking skills, and confidence. And when students are exposed to a classroom environment where good questions are the norm, something else wonderful happens too. The "let's try it and see what happens" thinking at the heart of any good question motivates students to generate their own good questions! In Question B, since there is a range of possible answers (e.g., $0 = (x - 3)^2 + 4$ or $4 = -17(x - 5)^2 + 3$), students might wonder:

- What is a generalization about all possible answers?
- How might my strategy or approach change if the givens of the problem changed—say, what if the equation had exactly one real solution?

The nature of the question stimulates students' natural sense of curiosity, which in turn fosters the higher-level thinking and the problem-solving skills that are essential to the work of doing mathematics in school and beyond.

Scenario 2: Trigonometric Ratios

Here's one more example of the power of a good question. RG, a tenth-grade geometry student, had just finished a unit on right triangles where she had been asked to find the measures of side lengths and angles of particular triangles. She was able to solve these problems using ratios between sides of 30-60-90 triangles and 45-45-90 triangles. With this in mind, the teacher gave RG the following good question:

> *Roya was told that two sides of a right triangle were 6 and 12. What might the triangle look like and what could be the measures of the other side and two acute angles?*

RG read the question several times, and exasperated, responded, "I can't do this! There isn't enough information." What happened?

To find an answer, RG needed to think about the relationship between the side lengths of a right triangle and its angle measures. She could reason that the triangle might be a 30-60-90 triangle (since one side is twice as long as the other) with a third side of length $6\sqrt{3}$ units. Or, she could reason that the legs measure 6 and 12 units and use the Pythagorean Theorem and trigonometric ratios to find the unknown side length and angle measures. This certainly required her to do more than remember a fact or reproduce a skill. It required higher-order reasoning skills involving comprehension of the task, application of the concepts and appropriate skills, and analysis and some synthesis of the major concepts involved. Through further probing, this question allowed the teacher to see that RG had little appreciation for the relationship between side lengths and angles of right triangles. She had learned to answer routine problems without fully understanding the concepts.

Feature 2: Students Can Learn by Answering Good Questions, and the Teacher Learns about Each Student from the Attempt

Good questions have the potential to make students more aware of what they do know and what they do not know. That is, students can become aware of where their understanding is incomplete as well as where they have developed misunderstandings. Let's revisit the earlier question from Scenario 1:

> *Tuhina looked at a quadratic equation in vertex form and thought to herself, "Based on the graph of the related function, I know the equation has no real solutions." What might be the equation Tuhina was looking at?*

The very act of trying to answer the question can help students gain a better understanding of the concepts involved. Imagine that a group of students told their teacher, "We know that when we solve a quadratic, if we get a step where *x* squared is equal to a negative number, the equation has no real solutions. But how can a graph help us see this?" By answering this question, the students and their teacher both learn about their current level of understanding of multiple representations of functions. The students understand how to use symbolic methods to solve quadratic equations. This is encouraging! However, they do not yet appreciate the value of a graph in revealing information about the solutions to a quadratic equation. With this revelation, the students and their teacher can now work toward helping them reach a deeper level of understanding so that the students can connect what they know about solutions derived symbolically to the intersection of two functions on a graph. Use of the good question not only reveals common incorrect responses to the teacher (proving to a be a valuable assessment tool) but can empower students to use their mistakes as learning tools.

Feature 3: Good Questions May Have Several Acceptable Answers

Many of the questions teachers ask, especially during mathematics lessons, have only one correct answer. Such questions are acceptable, but there are many other questions that are open-ended, allowing multiple answers, and teachers should make a point of asking these, too. Each of the good questions that we have already looked at has more than one possible answer. Because of this, these questions foster higher-level thinking, encouraging students to develop their problem-solving expertise at the same time as they are acquiring mathematical skills. Good questions also dispel the myth that math problems always have just one right answer.

Feature 4: Good Questions Are Accessible to All Students

Good questions invite all students into the conversation. Because good questions allow for multiple entry points, students can use their prior knowledge as a gateway toward new understandings. Take, for example, the earlier question asked of RG on the next page.

Roya was told that two sides of a right triangle were 6 and 12. What might the triangle look like and what could be the measures of the other side and two acute angles?

Some students may access this question by making a sketch while others may rely on the Pythagorean theorem or trigonometric ratios. Because this question has more than one answer, it supports students with emerging, developing, or advanced understanding of right triangles and trigonometric ratios. Good questions are accessible to all students because they do not require only one set pathway or approach.

Looking Ahead

In this chapter, we have looked more closely at the four main features that categorize good questions. The next chapter shares two ways to construct your own good questions.

CHAPTER 2

How Do I Create Good Questions?

When you first start using good questions, we suggest you select your questions from the collection of questions in Part II, "Good Questions to Use in Math Lessons." After a while you will want to create good questions for yourself. Detailed in this chapter are two methods that can be used to create good questions. The one you use is a matter of personal preference.

Two Methods for Creating Good Questions

1. Working Backward
2. Modifying a Standard Question

When creating your own good questions, it's important to plan the questions in advance, because creating them is not something that can be done in the moment. Creating good questions relies heavily on the destination we have in mind. What do we expect our students to be able to do, say, or understand by the end of the lesson? Beginning with the end in mind requires us to start with a clear understanding of desired knowledge, learning, and outcomes. When we think about questions that we might ask our students, it is helpful to first ask ourselves a few questions; for example:

- What are the mathematical goals of the lesson?
- What are the connections we'd like students to make between lesson goals and previously covered concepts and/or procedures?
- What are the common incorrect responses students may give?
- What is our assessment of students' understanding?

Method 1: Working Backward

This is a three-step process.

The Working Backward Method

Step 1. Identify a topic.
Step 2. Think of a closed question and write down the answer.
Step 3. Make up a question that includes (or addresses) the answer.

For example:

Step 1. The topic for tomorrow is rational functions.
Step 2. The closed question might be *What is the y-intercept, if any, of the graph of the rational function* $f(x) = \frac{x+6}{x}$? (Answer: The graph does not have a y-intercept.)
Step 3. The good question could be *Write the equation for a rational function that does not have a y-intercept.*

Some more examples of how this works are shown in the following table.

STEP 1 Identify a topic.	STEP 2 Think of a closed question and write down the answer.	STEP 3 Make up a question that includes the answer.
systems of linear equations	What is the solution to this system? $y = 8x - 3$ $y = 8x + 5$ (Answer: No solution.)	What is a possible second equation of the following system so that the system has no solution? Why? $y = 8x - 3$ ___?___
complex numbers	What is the sum of -8 + 3*i* and 1 + 2*i*? (Answer: -7 + 5*i*.)	If the sum of two complex numbers is -7 + 5*i*, what are possible values of the two numbers?

STEP 1 Identify a topic.	STEP 2 Think of a closed question and write down the answer.	STEP 3 Make up a question that includes the answer.
transformations	If the line $f(x) = x$ is translated down 3 units, what is the resulting function? (Answer: $f'(x) = x - 3$.)	Eddie took the line $f(x) = x$ and translated it down 3 units. Estelle instead translated the line $f(x) = x$ to the right 3 units. How are their resulting functions the same or different? Explain.

Method 2: Modifying a Standard Question

This is also a three-step process.

The Modifying a Standard Question Method

Step 1. Identify a topic.
Step 2. Think of a standard question.
Step 3. Modify it to make a good question.

For example:

Step 1. The topic for tomorrow is congruent properties of triangles.

Step 2. A typical exercise might be *Assume point* A *lies on some line l. If* ΔABC *is symmetric with respect to line l, which angles of* ΔABC *must be congruent?* (The answer is angles *B* and *C*.)

Step 3. The good question could be *Assume point* A *lies on some line l. If* ΔABC *is symmetric with respect to line l, what statements can you make about* ΔABC*?*

Some more examples of how this works are shown in the following table.

STEP 1 Identify a topic.	STEP 2 Think of a standard question.	STEP 3 Modify it to make a good question.
rational functions	What are the x-intercepts of the graph of the rational function whose equation is given below? $$y = \frac{x^2 - 4x}{x^2 + 5}$$	Write the equation for a rational function with x-intercepts (0, 0) and (4, 0).
trigonometric ratios	For right triangle *GQP*, $\sin Q = \frac{5}{13}$ and angle *P* is a right angle. Find cos *Q* and tan *Q*.	GQ drew a right triangle *GQP* and found that $\sin Q = \frac{5}{13}$. What can you determine about *GQP*?
probability	Consider a standard deck of cards. If *A* = the card is an Ace and *B* = the card is a face card, list the events that correspond to "not (*A* or *B*)."	Consider a standard deck of cards. Define two events *A* and *B* and list the events that correspond to "not (*A* or *B*)."

Looking Ahead

The more experience you have with good questions, the more you will want to use them, and the easier it will become for you to make up your own. Refer to either or both of the methods outlined in this chapter to help you feel confident in creating your own (in doing so, you might also "invent" additional methods!). In the next chapter, we outline five steps to successfully implementing good questions in your classroom.

CHAPTER 3

Five Steps to Successfully Implementing Good Questions in Your Classroom

Today's mathematics classrooms should be dynamic places where students are involved and engaged in their own learning. This can be achieved through activities that promote higher-level thinking, cooperative problem solving, and communication. We have seen that good questions support these activities and are readily available for teachers to use. But how can you ensure success with good questions? In this chapter we outline five steps we've found helpful in successfully implementing good questions, including advice for overcoming problems that might arise at each stage. We then take you through each of the steps using a specific good question.

Five Steps to Successfully Implementing Good Questions

1. Prepare the good question.
2. Present the good question.
3. Students work on the good question.
4. Facilitate a whole-class discussion.
5. Summarize.

Step 1: Prepare the Good Question

See Chapters 1 and 2 for guidance in understanding what makes up a good question as well as creating good questions. Remember that Part

II of this resource offers many classroom-tested good questions for you to select and use as well.

Once you've selected a good question, it's critical that you understand the mathematics embedded in the question. On your own or with colleagues, think through the question and solve it. This is of particular importance with good questions because they might have more than one answer and/or approach. Try to think of more than just one answer. Doing so will help you anticipate and then react to the variety of responses you will hear from your students (undoubtedly, students will think of things you did not!). Remember, working through a good math problem takes time. An understanding of the mathematics involved in the question will help you determine the amount of time students might need to process and answer it. Following are questions we've found helpful in working through a good question prior to teaching with it.

Questions for Preparing a Good Question

- Do I have the necessary materials (e.g., graph paper, chart paper)? If not, where can I get them?
- What common incorrect responses or difficulties might my students have with the language, concepts, or directions?
- What follow-up questions can I ask that will readdress or redirect common incorrect responses or difficulties?
- For groups that finish early, what follow-up questions can I ask to enrich the discussion?
- How much time will students need to answer the question?

Step 2: Present the Good Question

When presenting the good question to students, display it on the board where everyone can see it. As you pose the question, refer to the corresponding words on the board. Make sure that all students know what the question is; do not assume they know just because they can see it. Consider asking some students to repeat the question in their own words.

Allow time for students to ask you about the meaning of the question. Clarify the question to them if necessary but do not give any directions or suggestions on how to do it. This is for students to work out for themselves. The following is a summary of essentials we like to revisit before presenting good questions.

Essentials for Presenting Good Questions

- Present the question clearly using accessible mathematical language.
- Set clear and reasonable expectations for student work.
- Allow for individual approaches, methods, and/or answers.
- Ensure that materials and technology (e.g., online graphing software) are available for student use.

Step 3: Students Work on the Good Question

Once the question has been introduced, allow ample time for student investigation. We suggest having students work in pairs or small groups. This gives them the opportunity to communicate their ideas to others. Not only do students hear and learn from one another's answers, they also rehearse their thinking before sharing with the whole class (we've found that students are often more willing to share their reasoning and ideas with the whole class once they have first had a chance to practice with a small group). Encourage students to keep written records of their thinking together—calculations, charts, diagrams, and/or pictures. This documentation will help them support and defend their thinking and answers.

Pair or small-group work can also benefit students who may have difficulty starting. Students should be encouraged to go to each other first before seeking help from the teacher. Suggest that they try representing the problem in some way, such as by drawing a diagram or using materials.

If, as students are working, you find that there are too many who cannot make progress without your assistance, you might need to reconvene as a class. Have students share their concerns. If the concerns of each group—or individuals within each group—are all different, then this is a sign that the question you have posed may need to be further modified. You could also decide to abandon the question altogether as unsuitable at this stage. If this happens, do not worry, because it takes time and practice to choose appropriate good questions. We assure you it's worth the effort and perseverance.

Once students are working, monitor their progress and note strategies, partial understandings, and misconceptions. If students stop after giving one response, ask them to look for other possible answers. If they have found all possible answers, ask them to explain their thinking behind each answer. You could also ask a related question to extend their learning. For example, you could change the parameters of the problem, add an additional step, or change the size of the numbers in the problem.

Step 4: Facilitate a Whole-Class Discussion

Facilitate a whole-class sharing of answers, strategies, and the discovered mathematics. This is a meaningful time for students to reflect and build on their learning. Class discussions offer opportunities for students to achieve understanding by processing information, applying reasoning, hearing ideas from others, and connecting new thinking to what they already know (Chapin, O'Connor, and Anderson 2013).

Consider starting your discussion before all student groups have finished the task; it is better to stop while students are still engaged with the question and interested in the task. Consider signaling to groups that they have five minutes left before the whole-class discussion.

As students share, write their responses where everyone can see them and/or display their visual representation. If there are errors, work together to identify such and figure out why. Often, as students are explaining their work, they discover the error themselves.

Guide the conversation by keeping the focus on students' thinking. Encourage students to support, add to, and even disagree with one another's strategies. Have them address one another, not you. The following questions may help guide and facilitate discussions of students' answers.

Questions to Prompt Whole-Class Discussions

- Why do you think that?
- How did you know to try that strategy?
- How do you know that is the answer?
- Will this work with every number? Every similar situation?
- When will this strategy not work? What is a counterexample?
- Who has a different strategy?
- How is your answer similar to or different from another student's answer?
- Repeat your classmate's ideas in your own words.
- Do you agree or disagree with your classmate's idea? Why?

Step 5: Summarize

Typically, a summary of the main learning will naturally evolve as the whole-class discussion develops. Continually check for understanding; just because a few students are eager to share does not mean everyone understands. To wrap up, review key points with everyone. Reference visual representations and use teaching aids as needed to support the needs of your students. Relate the answers back to the question. It is also helpful to pose more questions using a similar format so that students can apply what they have learned to new situations.

The Five Steps in Action

Let's have a look at how the five steps might unfold in a high school mathematics class studying quadratic equations.

Step 1: Prepare the Good Question

Ms. Yancey wants a question to supplement the unit on quadratics that her students are learning. She selects the following question and solves it. With colleagues, she identifies several big ideas that she wants to target including the following:

- The graph of any equation of the form $y = a(x - h)^2 + k$ can be thought of as a transformation of the graph of $y = x^2$.
- The value of a in $y = a(x - h)^2 + k$ reveals whether the parabola has a minimum or maximum point.
- The ordered pair (h, k) reveals the location of the vertex of the related parabola.
- We can reason about the number of solutions or zeroes to an equation of the form $0 = a(x - h)^2 + k$ by sketching or visualizing the related parabola.

She notes to ask follow-up questions to target these ideas:

- Why is the equation $y = a(x - h)^2 + k$ referred to as vertex form?
- What do the values of a, h, and k tell you about the graph of the related function?

She notes that, if needed, she can make the question easier by asking students to use graphing software to graph quadratic equations in vertex form and asking students to connect particular constraints of each equation to the characteristics of the graph. Based on this preliminary work, she decides to allow approximately ten to fifteen minutes for students to investigate the question.

The Good Question

Tuhina looked at a quadratic equation in vertex form and thought to herself, "Based on the graph of the related function, I know the equation has no real solutions." What might be the equation Tuhina was looking at?

Step 2: Present the Good Question

Ms. Yancey writes the question on the board where all students can see it. She asks some students to read the question out loud and asks others to tell her what it means in their own words. She gives time for students to ask questions, being careful not to give any directions or suggestions on how to do the task.

Step 3: Students Work on the Good Question

Ms. Yancey has students work on the question in small groups. She circulates and observes. One group has written an equation, but tells Ms. Yancey, "we just took a guess." She asks this group to sketch the parabola of the related function and use this to check their answer. Another group stops after discovering one response. Again, Ms. Yancey encourages this group to sketch a graph and also encourages the group to look for other possible answers. Several other groups find a few answers; she asks them to think of a way to describe all their answers—is there a generalization? What is it? For one group that finishes more quickly than the others, she prompts them to form the generalization that any quadratic of the form $f(x) = a(x - h)^2 + k$ where a and k are both positive or both negative will have no real solutions because their graphs never touch the x-axis.

When all groups have at least one response to the question, Ms. Yancey asks each group to post possible equations—and sketches of related graphs—on the board. She asks them to talk in their groups about what they notice. She is OK with groups being at different stages in their work.

Step 4: Facilitate a Whole-Class Discussion

Ms. Yancey brings everyone together for a whole-class discussion. She asks students to share their observations. The answers from four different groups are:

- Group B: *All of the equations are in vertex form.*
- Group E: *For all of the graphs that open up, the vertex is above the* x-*axis.*
- Group C: *For all of the graphs that open down, the vertex is below the* x-*axis.*
- Group A: *For any quadratic of the form* $f(x) = a(x - h)^2 + k$ *where* a *and* k *are both positive or both negative will have no real solutions because their graphs never touch the* x-*axis.*

Ms. Yancey notes that these four responses differ not only in the level of mathematical understanding but also in the quality of thinking that is demonstrated by the answers. She takes a positive approach to each group's response, keeping the discussion focused on the relationship between the equation and the graph. She asks students to articulate the relationship between the vertex and the number of real solutions to the equation. She continually encourages discussion using questions like:

- "Why do you think that?"
- "How is your answer similar to or different from another student's answer?"
- "Do you agree or disagree with your classmate's idea? Why?"

Step 5: Summarize

Ms. Yancey reviews the main points: the relationship between the constraints of a quadratic equation in vertex form and the characteristics of the related graph, the values of a, h, and k in a quadratic equation of the form $0 = a(x - h)^2 + k$, can help us visualize or sketch the graph of the related quadratic function, and, in turn, this sketch can help us make statements about the solutions to the quadratic equation. Even though these points were brought up in the discussion, she revisits them. She pushes students' thinking, asking them, "Why is the equation $y = a(x - h)^2 + k$ referred to as *vertex form*? Which parameters tell you about the vertex of the related parabola? Which tell you about whether the parabola has a minimum or maximum point?" Through all of this, Ms. Yancey makes sure students do not lose sight of the original question.

Ms. Yancey poses a similar task:

What are the characteristics of the graph if a quadratic equation of the form $0 = a(x - h)^2 + k$ has exactly 1 solution or exactly 2 solutions?

Looking Ahead

Now that you have an understanding of what makes a good question (Chapter 1), methods for creating good questions (Chapter 2), and the basics for using good questions in your classroom (Chapter 3), we invite you to explore the second part of this resource, which is filled with our favorite good questions. We hope that you too will find them useful as you incorporate more questioning techniques into your classroom.

PART II

Good Questions to Use in Math Lessons

In Part II, we offer examples of good questions for you to select from and use in your classroom. Each question is accompanied by notes to support your teaching.

The Organization of Questions

The questions are categorized by mathematical topic. Many of the questions in these topics can be adapted to meet the needs of the students in your classroom by making them easier or more difficult.

In general, the questions in each section are neither hierarchical nor sequential. Each can be posed individually. Each section can be viewed as an à la carte menu of question choices to help students deepen their understanding of the mathematics they are studying.

The Investigations

As you are reading through the good questions, you will find some instances where they have been written as investigations rather than questions. This has been done where we felt they were better written as investigations. Use them in exactly the same way as the questions.

The List of Experiences

At the beginning of each topic is a list of experiences that students should encounter for the particular topic. Not all students will be ready for these experiences at the same time. It is quite possible that some students might be working on some of the experiences listed for preceding topics. They should not be treated as a progression of experiences but rather as a range of possible experiences.

The List of Materials

A list of materials that you might need, comprised of accompanying reproducibles, is also provided at the beginning of each topic. You will not need all of these materials unless you complete every question listed for the topic. Check that you have suitable materials before you present a question to your class.

The Teacher Notes

Following most questions there are teacher notes. Some identify important teaching points for the particular question while others describe in more depth the mathematics being addressed or common incorrect responses that may present themselves. Teaching notes may also present possible extensions or questions that can be assigned as homework. At other times they will be useful in helping you assess students so you can plan to overcome any difficulties. It is a good idea to make notes as you observe students working to use in future planning.

CHAPTER 4

Good Questions for Number and Quantity

As they progress through high school, the focus of students' work in mathematics extends beyond the set of rational numbers to include work with irrational numbers, all real numbers, imaginary numbers, and complex numbers. Students must make sense of the characteristics of individual sets of numbers as well as the relationships between overlapping sets (e.g., irrational numbers as a subset of real numbers). Carefully crafted questions can help support our students as they use sets of numbers to calculate, solve equations, or represent measurements.

Exponents

Experiences at This Level Will Help Students To

- simplify expressions involving exponents, including 0 and negative exponents and rational exponents
- use the fundamental law of exponents to compare and simplify expressions
- connect exponential and radical forms for rational exponents
- rewrite expressions involving radicals and rational exponents using the properties of exponents

Good Questions and Teacher Notes (pages 25–27)

1. Interpret the expression $25^{\frac{3}{2}}$ without using the terms *three halves* or *3 over 2*.

This question encourages students to think about the connections between rational exponents and radicals. For example, the expression could be interpreted as the square root of 25 cubed, which can be simplified using mental math: $(\sqrt{25})^3 = 125$.

2. What number might replace *x* to make a true inequality?

$2^9 + 2^9 < 2^x$

Since $2^9 + 2^9 = 2(2^9) = 2^{10}$, any value of x that is greater than 10 will satisfy this inequality.

3. Replace *a*, *b*, *c*, and *d* with integers to satisfy this equation.

$a^b \times c^d = 8$

One possible answer is the following:

$$(20^3)\ (10^{-3}) = 8$$

Encourage students to generate several possible answers and generalize about what they have in common. As an additional challenge, add the constraint that the integers must all be distinct.

4. Replace *a*, *b*, *c*, and *d* with distinct values (for *b*, *c*, and *d*) to satisfy this equation:

$$a^b \times c^d = a$$

One possible answer is given here:

$$3^{-5} \times 27^2 = \frac{3^6}{3^5} = 3$$

If students are struggling to engage with this question, you could scaffold by asking them to find values that satisfy $a^b \times a^d = a$. Then, ask them to find another way to express the value of a.

5. Chi observed that $2^7 + 2^7 = 2^8$. Chi asked, "Is this true for other bases? For example, does $3^7 + 3^7 = 3^8$?" How would you respond to Chi's question?

No, this statement is generally not true for most bases. It is true for powers of 2 because doubling can be achieved by adding a number to itself or multiplying a number by 2. For example, $2^7 + 2^7 = 2^8$ because $2^7 + 2^7 = 2(2^7) = 2^8$. However, $3^7 + 3^7 \neq 3^8$ because $3^7 + 3^7 = 2(3^7) \neq 3^8$.

As an extension, ask students to simplify the expression $3^7 + 3^7 + 3^7$. What do they notice? What generalizations can they make?

6. Write an expression of the form a^b where $a^b = \sqrt{2}$ and a is a positive integer not equal to 2.

If students are having trouble engaging with this task, first check to make sure that they know the equivalent relationship between $\sqrt{2}$ and $2^{\frac{1}{2}}$. Then, ask them to consider fractions equivalent to one-half. For example, one possible answer is $2^{\frac{1}{2}} = 2^{\frac{4}{8}} = 16^{\frac{1}{8}}$.

7. Replace the letters a and b with a positive value for a and a negative value for b in the expression a^b to create two different expressions such that one simplifies to a positive, rational number and the other to a positive, irrational number.

This question addresses the common incorrect response that negative exponents signal negative numbers. One possible set of expressions is $4^{(-\frac{1}{2})}$ and $2^{(-\frac{1}{2})}$.

Irrational Numbers

Experiences at This Level Will Help Students To

- define a fraction $\frac{a}{b}$ (where *a* and *b* are integers and *b* is not equal to 0) as a rational number that can be written as a decimal that either repeats or terminates
- classify every real number as either rational or irrational
- understand that irrational numbers can be represented with decimals that neither repeat nor terminate
- approximate irrational numbers using a terminating decimal

Good Questions and Teacher Notes (pages 29–31)

1. **What number might replace the question mark to create an expression that simplifies to a rational number?**

$\sqrt{11} \times \sqrt{?}$

Encourage students to form generalizations among their answers. For example, if the missing value is any perfect-square multiple of 11, the expression simplifies to a rational number (e.g., $\sqrt{11} \times \sqrt{44} = 22$). You can use this question to motivate a discussion about the following prompt: The product of two irrational numbers is (never, sometimes, or always) irrational.

2. **MJ's teacher asked him to determine whether the following number was rational or irrational: 3.212112111211112. . . . MJ said, "The number is rational because the decimal has a repeating pattern. Repeating decimals are rational numbers." Do you agree or disagree with MJ's answer? Explain.**

MJ's answer is not correct. The number does not have a repeating pattern because it does not have a repetend—or an identical string of digits—that repeats over and over again. The number is irrational.

3. Talisha used her calculator to write $\frac{7}{23}$ as a decimal. She entered "7 ÷ 23" and the number 0.304347826 appeared on her screen. Talisha said, "This tells me that seven twenty-thirds is an irrational number because this decimal does not repeat or terminate." Do you agree or disagree with Talisha's reasoning? Explain.

This question targets a common incorrect response about repeating decimals. Specifically, students can mistake a decimal with a long repetend for an irrational number. It is important to address this common incorrect response by revisiting the definition of a rational number. For example, $\frac{7}{23}$ is rational because it is written as a fraction of two integers. Therefore, it is a rational number.

4. Find an irrational number that is between 6.78 and 6.79.

You might also consider asking students to find a rational number between the two numbers. You can use this question to motivate a discussion about the density property of irrational numbers.

5. Find a rational number between 2.311311311311311. . . and 2.313113111311111. . . .

One possible rational number would be 2.312. You might also ask students to find one irrational number as well.

6. Create three numbers that satisfy the constraints described here:

0 < repeating decimal < terminating decimal < irrational number < 1

One possible answer is as follows:

$$0 < 0.\overline{002} < 0.16 < \sqrt{0.2} < 1$$

7. Write a list of exactly five numbers so that exactly three are rational, exactly two are integers, and exactly one can be represented by a repeating decimal.

This question helps students think about one number as a member of (at least) two sets of numbers. For example, in order to satisfy the constraints to the problem, one of the rational numbers must be a fraction that is equivalent to a repeating decimal (e.g., $\frac{1}{3}$).

Square and Cube Roots

Experiences at This Level Will Help Students To

- use square root and cube root symbols to represent solutions to equations
- evaluate square roots of small perfect squares and cube roots of small perfect cubes
- rewrite roots using exponents and apply the properties of exponents to simplify expressions involving roots
- apply square roots and cube roots to problems about geometric measurement

Good Questions and Teacher Notes (pages 33–37)

1. The solution to an algebraic equation is $x = 2\sqrt{3}$. What might have been the equation?

Two possible answers are given here:

$$2x = 4\sqrt{3}$$
$$x - \sqrt{3} = \sqrt{3}$$

Some students might suggest $x^2 = 12$. If so, be sure they understand that this equation has two solutions: $x = 2\sqrt{3}$ and $x = -2\sqrt{3}$.

2. Fabio used his calculator to simplify $\sqrt{1.44}$. He said, "Something must be wrong. It doesn't make sense that the square root of 1.44 is 1.2." Why might Fabio have said this? What might be confusing him?

Fabio might be confused that the square root of 1.44 is close in value to 1.44. This confusion might be related to a common incorrect response that a number's root must always be significantly smaller than the number. (In fact, some students think that the square root of a number is always half that number.)

Fabio can gain clarity by reminding himself that 1.2×1.2 can be interpreted as taking one and a little bit more groups of 1.2, so the product will only be slightly bigger than 1.2. Or, he can reason as follows:

$$1.44 = \frac{144}{100}; \sqrt{1.44} = \sqrt{\frac{144}{100}} = \frac{\sqrt{144}}{\sqrt{100}} = \frac{12}{10} = 1.2$$

3. Find values for a and b such that $a \neq b$ and $\sqrt{a} \times \sqrt{b}$ simplifies to a whole number value.

One possible answer is $\sqrt{3} \cdot \sqrt{12} = \sqrt{3} \cdot \sqrt{3} \cdot \sqrt{4} = 3 \cdot 2 = 6$.

If students constrain their answers to only perfect-square values for a and b, add the constraint "where a and b are not perfect squares" so that they consider other possibilities as well.

4. Find values for a and b such that $\sqrt{a}$ and $\sqrt[3]{b}$ are whole numbers that satisfy the inequality $\sqrt{a} < \sqrt[3]{b}$.

One possible answer is $\sqrt{16} < \sqrt[3]{125}$.

5. Find values that satisfy the inequality $1 < \sqrt{a} < \sqrt[3]{b} < \sqrt{c} < 4$.

One possible answer is as follows: $1 < \sqrt{2} < \sqrt[3]{8} < \sqrt{9} < 4$

As an extension, change the given interval from (1, 4) to (0, 1). (One possible answer: $0 < \sqrt{0.80} < \sqrt[3]{0.729} < \sqrt{0.9} < 1$.)

6. Val looked at the number $\sqrt{5.76}$ and said, "This must be irrational because 5.76 is not a whole number." Do you agree or disagree with Val's reasoning?

Some students hold the misconception that square roots of decimal numbers are always irrational. Students can use the square root key to find that the number shown is rational. As a follow-up, students can find other examples of square roots that are not integers but are still rational.

Rewriting decimals as fractions can also help to unravel this common incorrect response: $\sqrt{5.76} = \sqrt{\frac{576}{100}} = \frac{\sqrt{576}}{\sqrt{100}} = \frac{24}{10} = 2.4$.

7. Is it ever possible for the cube root of a number, n, to be greater than the square root of n? Explain.

Students may wish to rewrite the conditions of the question in symbols: Is it ever true that $\sqrt[3]{n} > \sqrt{n}$?

This question encourages students to revisit key ideas about the size of the factors and product when multiplying fractions less than one. One possible way for students to answer this question is to take examples from different classes of numbers including fractions less than one, fractions greater than one, and whole numbers. Students can collect and sort their examples to find that yes, the cube root of a number is greater than its square root when the number is a positive fraction less than one (e.g., $\sqrt[3]{0.729} > \sqrt{0.729}$).

8. NB looked at a radical expression and said, "I can tell just by looking at this expression that it is not in its simplest form." Find a radical expression that meets this description.

NB might have been looking at any square root that contains an easily recognized square factor (e.g., $\sqrt{4}$, $\sqrt{12}$, or $\sqrt{27}$). Or, NB might have been looking at a radical expression with a radical in its denominator (e.g., $\frac{2}{\sqrt{2}}$).

9. The sum of two radical expressions simplifies to $5\sqrt{3}$. What might be the expressions?

Post students' answers (e.g., $2\sqrt{3} + 3\sqrt{3}$). After verifying each answer, ask students to generalize. Do students see this as an application of the distributive property (e.g., $a\sqrt{3} + b\sqrt{3} = (a + b)\sqrt{3}$)?

10. A cylinder has a volume that is 21π cubic units. What might be its radius and height?

One possible answer is $r = \sqrt{3}$ and $h = 7$;
$V = \pi r^2 h = \pi(\sqrt{3})^2(7) = 21\pi$.

Students sometimes struggle to interpret square roots—especially when they are used to represent values in measurement formulas. For example, they may interpret a root within an expression for volume of a cylinder (e.g., $\pi\sqrt{3}^2 h$) as a statement that the radius is 3. Once students come up with possible answers to the question, make sure they can interpret—and approximate—the values expressed as roots.

11. **Rewrite each statement using exponents to show whether each statement is true or false for all positive rational values of x:**

Statement 1: $\sqrt{x} + \sqrt{x} = x$

Statement 2: $\sqrt{x} \cdot \sqrt{x} = x$

Statement 3: $\sqrt{x} \cdot \sqrt{y} = \sqrt{xy}$

Statement 4: $\frac{a}{\sqrt{a}} = \sqrt{a}$

Possible answers are:

Statement 1 is false: $x^{\frac{1}{2}} + x^{\frac{1}{2}} = 2x^{\frac{1}{2}} \neq x$

Statement 2 is true: $\sqrt{x} \cdot \sqrt{x} = x^{\frac{1}{2}} \cdot x^{\frac{1}{2}} = x^1 = 1$

Statement 3 is true: $\sqrt{x} \cdot \sqrt{x} = x^{\frac{1}{2}} \cdot y^{\frac{1}{2}} = (x \cdot y)^{\frac{1}{2}} = \sqrt{xy}$

Statement 4 is true: $\frac{a}{\sqrt{a}} = a^1 \div a^{\frac{1}{2}} = a^{\frac{1}{2}} = \sqrt{a}$

In addition to sharing students' answers, be sure to also talk about why rewriting square roots using exponents can be helpful in forming generalizations.

12. **Cylinder A has the same height but twice the volume of Cylinder B. What might be the radius of each cylinder?**

You may want to post students' answers and then talk about what all answers have in common. For example, students should notice that for all answers, the radius of Cylinder A is $\sqrt{2}$ times the radius of Cylinder B. You may also ask them to write a general form for all possible answers: Given that Cylinder A has the same height but twice the volume of Cylinder B, if the radius of Cylinder B is r and its height is h (and its volume is $\pi r^2 h$), then the radius of Cylinder A is $r\sqrt{2}$; $\pi(r\sqrt{2})^2(h) = \pi 2r^2 h$.

Imaginary Numbers

Experiences at This Level Will Help Students To

- perform arithmetic operations with complex numbers
- represent complex numbers and their operations on the complex plane
- use complex numbers in polynomial identities and equations

Good Questions and Teacher Notes (pages 39–44)

1. If the sum of two complex numbers is $-7 + 5i$, what are possible values of the two numbers?

This problem allows students to practice the skills of adding and subtracting complex numbers beyond rote examples since they must think about the structure. A possible answer would be $2 + 3i$ and $-9 + 2i$.

2. Jas divided two complex numbers and got $-7 + 5i$ as a result. What might the two complex numbers have been? Is there a solution where the real and imaginary parts are integers?

This problem allows students to practice the skills of multiplying and dividing complex numbers beyond rote examples since they must think about the structure. A possible answer is to let the divisor be $3 - 4i$, then use multiplication to show the dividend must be $-1 + 43i$. This provides students with a method for practicing and checking their division of complex numbers skills. You could also change "Jas divided" to "Jas multiplied" in the question which practices the same skills differently.

3. Is it possible to add two complex numbers and get a sum that is *not* a complex number? What about a difference? Product? Quotient?

This problem allows students to explore the closure of the complex numbers under the four arithmetic operations. They should discuss whether a real number (or an imaginary number) is also considered a complex number (e.g., you cannot divide by $0 + 0i$). Are there exceptions?

4. Able was practicing creating and multiplying pairs of complex numbers, when he discovered a pair whose product was a real number. Brin confirmed Able's result and then claimed to have found a pair whose product was imaginary. What complex numbers might Brin have used to produce the two "surprising" results?

This problem will lead students to the complex conjugate by using differences of squares to eliminate the imaginary (*bi*) part. The follow-up asks them to figure out how to "eliminate" the real part. Ask students to generalize the results.

5. Find a polynomial of least degree with integer coefficients that has *at least* one real root and *at least* one complex root.

A possible solution would be $(x^2 + 1)(x - 1) = x^3 - x^2 + x - 1 = 0$.

Students should be prepared to discuss the process by which they arrived at their solutions.

This question could be used to assess whether students understand that complex roots come in conjugate pairs and also allows a discussion of linear and quadratic factors.

6. Consider a complex number $z = a + bi$. How are z, $i \times z$, $i^2 \times z$, $i^3 \times z, \ldots$ related in the complex plane?

You may want to encourage students to use small values for a and b. Also, ask students to discuss how the pattern will continue. They should notice that $i^4 \times z = z$ to remind them that $i^4 = 1$. This question encourages students to relate the arithmetic of complex numbers to the graphical connection of 90-degree rotations in the complex plane.

7. Suppose you are given a complex number z. For each separate scenario, if possible, find another complex number, such that:

- The sum is real.
- The difference is real.
- The product is real.
- The quotient is real.
- The sum is imaginary.
- The difference is imaginary.
- The product is imaginary.
- The quotient is imaginary.

If not possible, explain why.

All are possible. You may want to scaffold the question by telling students $z = z_1 = a + bi$ and the other complex number is $z_2 = c + di$. Then possible solutions would be:

- $2 + 5i$ and $6 - 5i$ (i.e., b and d are opposites, a and c are irrelevant)
- $2 + 5i$ and $6 + 5i$ (i.e., b and d are equal, a and c are irrelevant)
- $2 + 5i$ and $2 - 5i$ (i.e., they are conjugates; a and c are equal, b and d are opposites)
- $2 + 5i$ and $6 + 15i$ (i.e., only when z_2 is a real multiple of z_1)
- $2 + 5i$ and $-2 - 3i$ (i.e., a and c are opposites, b and d are *not* opposites)
- $2 + 5i$ and $2 - 5i$ (i.e., a and c are equal, b and d are *not* equal)
- $2 + 5i$ and $15 + 6i$ (i.e., c is a multiple of b and d is the same multiple of a; $\frac{c}{b} = \frac{d}{a}$)
- $2 + 5i$ and $5 - 2i$ (i.e., d is the opposite of a, and $c = b$)

Some possible extension questions are: What are some general patterns that arise? What are some exceptions?

8. Suppose that $2 - 3i$ and $1 - i$ are roots to a polynomial equation with integer coefficients. What must be true about the polynomial? What might be true? What must be false?

This question requires students to apply the concept that complex roots come in conjugate pairs and allows the students to expand the set of roots in a variety of ways.

Students should notice that the minimum degree is 4. It might be true that there is at least one real root.

9. Consider the complex number $z = 0 + 1 \cdot i$ or $z = i$.
What do you notice about $z, z^2, z^3, z^4, \ldots$?
Are there any other complex numbers that have this property?

This problem allows students to see the cyclical nature of the powers of i, but the second part requires that they consider patterns that recur that have different periods (e.g., $z = -1$ has a pattern of length 2 while $z = \frac{\sqrt{2}}{2} + i\frac{\sqrt{2}}{2}$ has a pattern of length 8). Also, students should be encouraged to map these graphically to see the geometric pattern which they may see later with de Moivre's theorem.

10. In class, Ky states that "the *sum* of a complex number and its conjugate is a real number." Daisy says, "No, I think it's the *product*." Who is right? Can you give an example?

Both students are correct. Ask students to generalize the results and also extend the thinking to the quotient and difference.

11. **Can you find the square root of a complex number? If yes, create an example. If not, explain why.**

The general idea is if $\sqrt{a + bi} = c + di$, then $a + bi = (c + di)^2$, and so $a = c^2 - d^2$ and $b = 2cd$. Thus an example that is solvable is $\sqrt{8 + 6i} = 3 + i$. Students may notice that $(-3 - i)^2 = 8 + 6i$, which is another solution to the equation $z^2 = 8 + 6i$, but different than $\sqrt{8 + 6i}$ in the same manner that $\sqrt{25} = 5$, but the solution to $x^2 = 25$ is ± 5.

Have students discuss how they created their example. Ask students to explain whether or not the answer is unique. This problem can be expanded to cover cube roots and beyond. This question can motivate the usefulness of polar form and de Moivre's theorem.

12. **Geneva says that addition of complex numbers creates a parallelogram when looked at graphically in the complex plane. Can you create an example that demonstrates this?**

The parallelogram is formed by the points representing the origin, each of the two complex numbers and the sum. You can also ask students to demonstrate this for subtraction, multiplication, and division.

13. Find three complex numbers z_1, z_2, and z_3 such that $z_1 = z_2^2 = z_3^3$?

Here, students might find the polar form easier to work with to find a solution. One possible solution is $z_1 = 64i$; $z_2 = 8\cos(\frac{5\pi}{4}) = -4\sqrt{2} - 4\sqrt{2}i$; and $z_3 = 4\cos(\frac{5\pi}{6}) = -2\sqrt{3} + 2i$.

Students should notice for any z_1 that there are six possible sets of z_2 and z_3. Ask the students to explain why.

14. Ky says that multiplying complex numbers is easier when they are in polar form rather than rectangular form. Jas agrees and adds that the opposite is true for addition and subtraction. Create an example or two to demonstrate their reasoning.

This allows students to represent complex numbers in both rectangular and polar forms and draw geometric connections.

15. If $\overline{z}$ represents the *complex conjugate* of z, then $\overline{z_1 + z_2}$ would represent the conjugate of the sum of two complex numbers. Is the conjugate operation commutative, associative, and/or distributive under the operations of addition, subtraction, multiplication, division, and/or exponentiation? For example, answering the question "does $\overline{z_1 + z_2} = \overline{z_1} + \overline{z_2}$?" would show that the operation distributes over addition.

Ask students to create examples (numerically and graphically) for each truth. This can be a good activity to split up among the class.

Note: This problem is adapted from *Advanced Mathematics* (Brown 1997).

Infinity and Limits

Experiences at This Level Will Help Students To

- understand that infinity is "uncountable" and undefined
- distinguish between the concept of infinity as a number and other situations where students encounter infinity, such as how non-repeating decimals have an infinite number of digits but that the value of the number is itself not infinity

Good Questions and Teacher Notes (pages 46–48)

1. Kaitlyn claims that she knows a value of x (which she calls k) that will result in the greatest output of $y = x^3$. Is this possible? Why or why not?

This question has students quickly recognize the nature of end behavior. In a discussion have students consider a proof by contradiction. Assume that $f(k) = k^3$ is the greatest output. Then $f(k + 1)$ should be less than $f(k)$. But $f(k + 1) = (k + 1)^3 = k^3 + 3k^2 + 3k + 1$. Since $k^3 + 3k^2 + 3k + 1 > k^3$ then $f(k + 1) > f(k)$, which is a contradiction. Therefore, there cannot be a value $x = k$ for which $f(k) = k^3$ is the greatest output.

2. What is 0.9999^∞? Why?

Some students will argue that 0.9999 is so close to 1, and since $1^\infty = 1$, then $0.9999^\infty = 1$. But, in fact, as $x \to \infty$, $0.9999^x \to 0$. One way to help students see this is to observe the graph of $f(x) = 0.9999^x$ and to look out at values where $x = 9{,}000$ and $x = 10{,}000$ to see how $f(x)$ is decreasing toward 0. In addition, have students consider what happens to a simpler case, such as 0.9^x. For example, when $\frac{9}{10}$ is squared, students should see how the numerator and denominator diverge ($\frac{81}{100}$), and this trend only continues ($\frac{729}{1{,}000}$, $\frac{6{,}561}{10{,}000}$, etc.).

3. Why can't numbers be divided by 0? How can the function $f(x) = \frac{1}{b}$ help explain?

Although students may state that dividing by 0 is not possible, few can explain why. This question encourages students to use a basic function to make sense of rules that they may have learned years earlier. As $x \to 0$ from the right, $f(x) \to \infty$. Yet when $x \to 0$ from the left, $f(x) \to -\infty$. So when $x = 0$, we cannot describe $\frac{1}{x}$ as either ∞ or $-\infty$.

4. Zeno figures he shouldn't bother to leave the classroom to head to his next class down the hallway because he suspects he'll never get there. Here's his logic:

To get to his next class, he has to pass the point halfway between his current classroom and his next class (referred to as *M*). However, to get to *M*, he must pass the point halfway between his current classroom and *M* (referred to as *M′*). To get to *M′*, he must similarly have to pass the midpoint between his starting point and *M′*(*M″*), and so on. Since to get anywhere, there is an infinite number of these halfway points to pass through, Zeno figures not only will he never reach his destination, but that he cannot even move.

What do you think? Explain your thinking.

This prompt gets students to think about countability. A number line with his starting and ending points can help students organize their thoughts. As they recognize that there are an infinite number of midpoints that Zeno passes through, they confront the fact that there are infinite points between any two points. Thus, any movement between any two points automatically consists of passing through infinite points. Of course, using this method to plan movement between two points would be problematic. However, Zeno does not need to assume that passing through infinite points limits movement.

5. Assume a regular polygon has n sides and has a radius (the length from the center of the polygon to each of its vertices) of 1 unit. What is the area of the polygon if $n \rightarrow \infty$? Explain.

This question can be asked in a geometry class to start developing students' concept of limits. As the number of sides of a regular polygon increases, the polygon increasingly approximates a circle. Therefore, as $n \rightarrow \infty$, the area approaches $\pi r^2 = \pi(1)^2 = \pi$ square units.

6. Anthony proposes that when written as a decimal, π goes to infinity because the decimal keeps going on and on without end. What do you think?

This prompt raises an issue that can be very confusing to students. On the one hand, students might recognize that as they increase the number of decimal places of π (such as 3.14 to 3.141 to 3.1415, etc.), each new extension of the decimal increases the value of the number. Therefore, carrying out this logic, since students are told that π has infinite non-repeating decimals, they may assume that π must increase to infinity. Yet the question "but then how can π be greater than four if its ones digit is three?" will help students to recognize the difference between the digits of π (of which there are infinitely many) and the value of the number of π (which is between 3 and 4).

CHAPTER 5

Good Questions for Symbolic Algebra

In secondary school, students extend elementary work with variables as unknowns (for example, solving for missing values) to include the idea of parameters (in other words, place holders for numbers such as *m* and *b*). Students are introduced to algebraic objects, such as variable expressions, which represent generalized quantities and lay the foundation for later work with functions. Students also develop strategies to manipulate variable expressions while maintaining equivalence, such as through combining like terms, factoring, and other simplification strategies. Moreover, relationships between variable expressions are explored (for example, under what conditions are two expressions equal?), leading to the development of solving strategies for equations and inequalities. Finally, this study extends to the solutions of systems of equations and inequalities.

Solving Equations

Experiences at This Level Will Help Students To

- rewrite equations in equivalent forms as a strategy for solving for an unknown
- understand that equivalent equations have the same solutions
- identify unknowns even when they are not called *x*, or even when there is an *x* in the equation
- understand that when equations are equivalent, they have the same solutions
- recognize that some equations can have no solutions
- define a solution to an equation as a set of numbers that make that equation true

Reproducibles are available in a downloadable, printable format. See page xx for directions about how to access them.

Materials

- Equations Venn Diagram (Reproducible 1)

Good Questions and Teacher Notes (pages 51–58)

1. How can we be sure that $x = 3$ is the only possible solution to $3x - 2 = x + 4$?

This question leads students to think beyond the solution to its uniqueness. One way to explore this question is to examine the idea of "solution of a linear equation" graphically. That is, when students visualize the linear equation as two functions that for some x can have equal y-values, then the uniqueness of the solution can be justified by the fact that two non-coinciding lines can intersect, at most, one time.

2. What equations have a solution of $x = -2$? How do you know?

Students may approach this via guess and check. However, this question provides an opportunity for students to think broadly of how equations can be "complicated" using operations that maintain equivalency. Recognizing this can help students see that the "solution" $x = -2$ is actually an equation that is equivalent to all equations that have only -2 as a solution.

3. How can you tell if two equations have the same solution(s)? Is there more than one way?

This question is intended to get beyond the obvious ("solve both and see if you get the same answer") to the recognition that if one can be manipulated via algebraic properties to become the other, they are equivalent and must have the same solutions.

4. Fill in the blank with *never*, *sometimes*, or *always*:

The expressions $\sqrt{x^2}$ and x are _____ equal.

The expressions $\sqrt{x^2}$ and x are sometimes equal. They are equal for values of x greater than or equal to 0. Note that we use the word *equal* here and not *equivalent* since *equivalent* conveys identical solution sets.

5. While solving $4(2x - 16) = 24$, a group of students came up with multiple different ways to rewrite the equation. Explain how each version came from the original. What are the benefits or limitations of each rewritten equation?

a. $2x - 16 = 6$

b. $4(x - 8) = 12$

c. $8x - 64 = 24$

d. $8(x - 8) = 24$

This problem draws attention to multiple strategies that can be used to solve equations.

6. If $x = 5$ is a solution to the equation $4(x + k) - 3x = 1$, then what must be the value of k?

This type of equation challenges students to recognize that the unknown does not need to be called "x" and sets up other types of problems later where students need to solve for a parameter.

7. What do the expressions here have in common? Write another expression that has the same trait.

$$x^4 - 9x^2 \text{ and } 4x^2 - 36$$

This question encourages students to look for similar structures and common factors to enable them to learn how to recognize structures that can be useful in later simplification and factoring problems. Both expressions have a $x^2 - 9$ factor. However, there are a lot of ways students could create new expressions that also have this factor, including splitting the $(x + 3)$ and $(x - 3)$ terms into new terms, such as with $(4x + 12)(2x - 6)$.

8. Generate at least two equations that have the same solution(s) as $2x^2 - \frac{2x}{3} = -18$.

These questions use the notion of equivalence to help students gain insight into the mathematical structure of equations. Strategies include multiplying everything by 3 to eliminate the fractions or dividing everything by 2 to eliminate the coefficient on the x^2 term.

9. For what x value(s) is the value of the expression $3x - 4$ the same as the value of the expression $10 - 2x$?

Although this question is essentially asking students to solve for x when $3x - 4 = 10 - 2x$, the different phrasing of this question calls to question what equations inherently represent, namely, a condition under which two expressions are equal. In addition, this formulation of the task could spur students to consider a functional approach to the question, considering when the system $y = 3x - 4$ and $y = 10 - 2x$ intersect. A discussion could compare these two approaches for how they are the same or different.

10. What is an equation that has no possible solution? Explain how you know your equation has no solution.

This question requires students to confront the meaning of equality. That is, to write an equation that has no solution, the students need to craft two expressions that cannot possibly be equal. For example, an absolute value can never be negative, so any equation for which the absolute value of an expression equals a negative number would have no solution. An equation of two expressions that each represent a pair of parallel lines (e.g., $3x - 4 = 3x + 10$) will also have no solution.

11. What equations could go in each part of this Venn Diagram? Write an equation for each region (including the region outside both circles).

(See Reproducible 1, Equations Venn Diagram.)

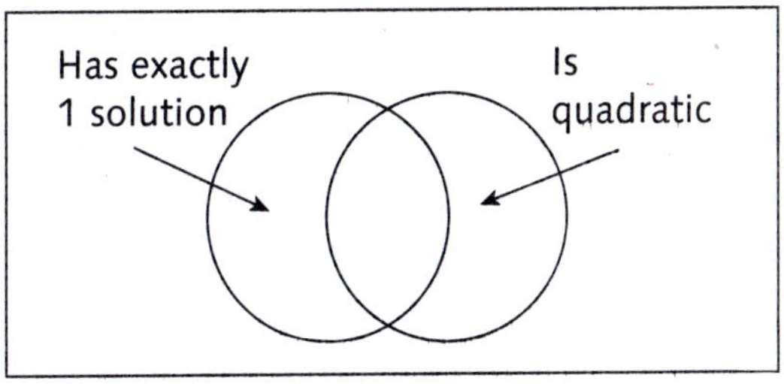

This question encourages students to look at equations as representing multiple categories that distinguish them from others.

12. What equations could go in each category in the following table?

Never True	Sometimes True	Always True

This question enables students to compare the qualities of equations that have different types of solutions (none, some, all).

13. Eileen started the following equation. How could you finish it so that it has no solution? Has only two solutions? Has infinite solutions? Try to come up with more than one answer for each.

$$-x + 5 =$$

This question encourages students to use the structure of an equation to reason about its number of solutions. Students might find it helpful to connect their equations to their related graphs. For example, we can determine that the equation $-x + 5 = -x - 10$ has no solution if we think about the equation in terms of two parallel lines. Since the solution represents the point of intersection, the equation has no solution.

14. Without solving the equation, Sadiki wants to know if $x = 4.5$ is a solution. How can he figure it out? Explain.

$$3 + 2(2x + 3) = 36 - 2x$$

Many students see checking a solution as a part of a solving process instead of a test of equality. Even students who are great at following a solving process and checking a solution will surprisingly be stumped by this question, which essentially removes the checking process from the solving procedure.

15. If $x = 4$ is a solution to $3(2x - 5) = 9$, then what sequence of equivalent expressions can change $x = 4$ to $3(2x - 5) = 9$?

This basically reverses the process of solving by repeatedly writing equivalent equations to find a solution to instead repeatedly writing equivalent equations to complicate the solution to build an equation. It helps students look beyond the solving procedure to recognize a chain of equivalent equations that connect a solution to its original equation. Although some students might realize they could solve $3(2x - 5) = 9$ for x and then just reverse the steps, other students may recognize that complicating x to build the left expression $3(2x - 5)$ can be done by seeing the order of operations. That is, double both sides to get $2x = 8$, then subtract five from both sides to get $2x - 5 = 3$, and finally multiply both sides by 3 to get $3(2x - 5) = 9$.

16. Can an equation have exactly three solutions? If so, give an example. If not, explain why not.

After students are introduced to solving quadratic equations, where equations can have up to two solutions, this question can raise the possibility for polynomials of higher order even before students have ever seen one before. That is, if $(ax - b)(cx - d) = 0$ has exactly two solutions, then $(ax - b)(cx - d)(ex - f) = 0$ can have exactly three solutions (under certain conditions), and so forth.

17. What can you say about the value of k so that the equation below has no solution? One solution? Two solutions? For each, justify your conclusion.

$$(2x - 1)^2 = k$$

This question focuses on the number of solutions, which places attention on the conditions of a perfect square. Since squaring results in a positive value or 0, then the number of solutions depends on whether k is positive (two solutions), 0 (one solution), or negative (no solutions).

18. Based on the equations here, which is greater: A or B? Explain how you know.

$$3x + 5 = A + x + 2 \text{ and } B(x - 3) = 2x^2 - 2x - 12$$

In this problem, the "solutions" for A and B do not resemble typical solutions, for which students usually get a numerical value. However, the resulting expressions can still be compared, as one is $2x + 3$ and the other is $2x + 4$. Since $2x + 4$ is always greater than $2x + 3$, B is greater.

Equivalence

Experiences at This Level Will Help Students To

- understand that expressions are equivalent if, for the given variables, the value does not change
- decide if two expressions are equivalent by representing both geometrically
- understand that expressions *A* and *B* are only equivalent if they are always equal

Reproducibles are available in a downloadable, printable format. See page xx for directions about how to access them.

Materials

- Area Diagram A (Reproducible 2)
- Area Diagram B (Reproducible 3)
- Graph of Expressions *A* and *B* (Reproducible 4)

Good Questions and Teacher Notes (pages 60–63)

1. A friend of yours wants to know the reasoning behind the following statement. Convince her that this statement is true, or, if it is false, explain why and adjust the expression on the right so that the statement is true.

$$\frac{x+1}{2x+1} \neq \frac{1}{2}$$

This question focuses on the simplification of rational expressions by drawing attention to a common error.

2. Assume all the angles in the following diagram are right angles. Explain how this diagram shows that $3(x + 2) = 3x + 6$.

(See Reproducible 2, Area Diagram A.)

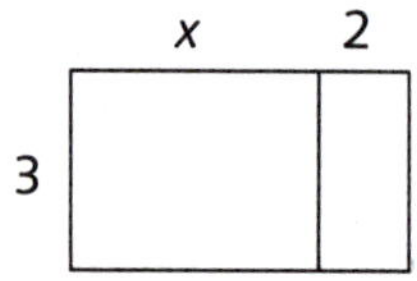

This problem requires students to see the sum of the areas of the two smaller rectangles (which are $3x$ and 6) as equivalent to the area of the whole rectangle (which can be written as $3(x + 2)$).

3. Draw a geometric diagram that explains why $x^2 + 6x + 14 = (x + 3)^2 + 5$.

This question can help students recognize that geometric representations of expressions can support the demonstration of equivalence. This question also lays a foundation for students to understand how to rewrite quadratic expressions by completing the square.

4. Assume all the angles in the following diagram are right angles. Explain how the diagram shows that $x^2 - 1 = x(x - 1) + x - 1$.

(See Reproducible 3, Area Diagram B.)

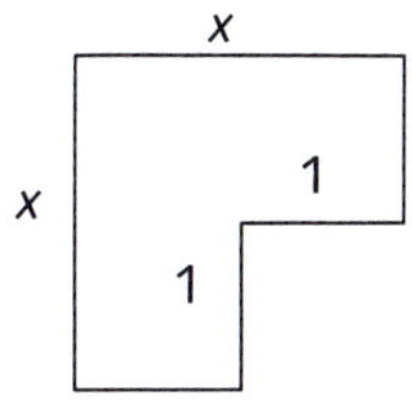

This question requires students to connect geometric reasoning with symbolic expressions. The left side of the identity represents the entire area found when an x-by-x square has a 1-by-1 square removed. The right side of the identity represents the sum of areas when the figure is broken into two rectangles.

5. Two expressions are equivalent if they are always equal. What expressions are equivalent to $(3x - 6)(x + 4)$? Try to generate at least three equivalent expressions.

This question encourages students to think flexibly about conditions for equivalence. Some students might notice the common factor in $3x - 6$ that allows an equivalence to be found through factoring: $3(x - 2)(x + 4)$. Some might note that this common factor could instead be multiplied into the $(x + 4)$ part of the expression, leading to $(x - 2)(3x + 12)$. Still others might multiply the expressions to find $3x^2 + 6x - 24$ or variations of this expression.

6. **In the following graph, line *A* is the graph of an expression *A* and line *B* is the graph of an expression *B*. What would need to be done to expression *A* so that it would be equivalent to expression *B*? Explain.**

(See Reproducible 4, Graph of Expressions *A* and *B*.)

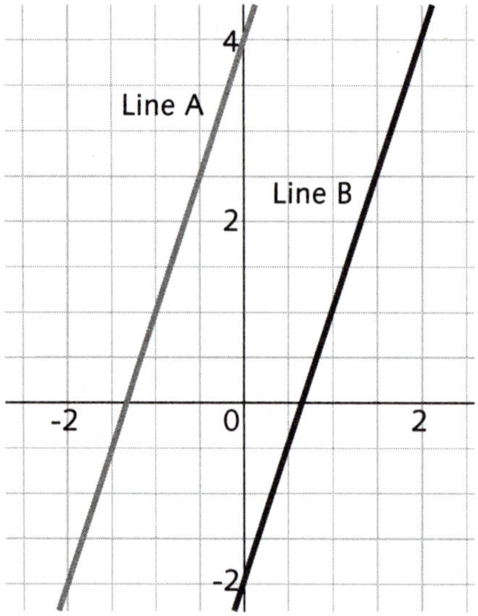

This question draws attention to the fact that expressions represent functions. That is, as the independent variable changes in an expression, the resulting value of the expression (the output) changes as well. This question also reinforces that equivalence requires the output of the two expressions to always be equal. Finally, this question draws on the notion of translation of functions. Students could recognize that the expression *A* could be translated down 6 units by subtracting 6 from the output of *A* or it could be translated to the right 2 units (adding 2 to the input of *A*).

7. If $x + 3 = -2$, then does $x + 4 = -1$? Why or why not?

This question encourages students to first compare these two equations for differences. Some students will perhaps solve both and find out their solutions are equal (i.e., $x = -5$ for both). However, this question also begs the issue that if $a = b$, then $a + 1 = b + 1$. Or, more generally, if $a = b$, then $a + c = b + c$. A follow-up equivalent equation could be found by adding or subtracting any amount from both sides.

8. Are any of these expressions equivalent? How do you know? Use an area model to justify your conclusion.

$3x + 6$ $\quad$ $9x$ $\quad$ $3(x + 2)$ $\quad$ $3 + 6x$ $\quad$ $3x + 6x$

This question can lead to a discussion about what terms can be combined. When students are required to represent with an area model, the dimensions of $3x + 6$ and $3 + 6x$ are not necessarily the same (i.e., they are only the same with $x = 1$).

Systems

Experiences at This Level Will Help Students To

- understand that a solution to a system of equations must make all equations true simultaneously
- use substitution and elimination methods to generate a one-variable equation that can be solved for a variable
- connect the graph of two functions to the number of solutions of the related system of equations (e.g., When the graphs of two functions never intersect, the system of equations has no solution)
- understand that a system of two linear equations can have 0, 1, or infinite solutions
- understand that a system of a linear equation and a quadratic equation can have 0, 1, or 2 solutions

Reproducibles are available in a downloadable, printable format. See page xx for directions about how to access them.

Materials

- Points (x, y) Venn Diagram (Reproducible 5)
- Annie's Graph (Reproducible 6)
- Graph of Transformed Functions (Reproducible 7)

Good Questions and Teacher Notes (pages 65–72)

1. What points (x, y) could go in each part of this Venn Diagram? Write at least one point (x, y) for each region (including the region outside both circles).

(See Reproducible 5, Points (x, y) Venn Diagram.)

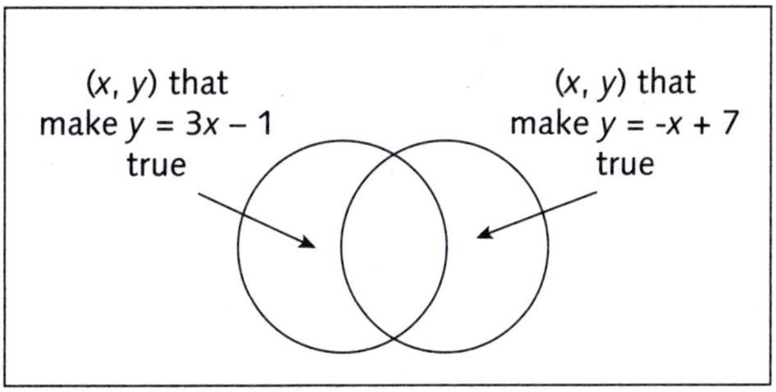

This question broadens the meaning of "solution" of an equation to a coordinate (x, y) and lays a foundation for the meaning of systems of equations.

2. Is (1, -4) a solution to this system of equations? How can you tell without solving?

$$3x + y = -1$$

$$2x - 3y = 12$$

This question requires students to test that the point makes both equations true. Some may stop after testing the top equation and determine the system is true because it makes the top equation true. However, a discussion can help these students understand that a solution to a system must make both equations true.

3. What systems of quadratic equations could go in each category in the following table?

No Solution	One Solution	Two Solutions	Infinite Solutions

This question enables students to compare the qualities of systems of equations that have different types of solutions (none, some, all). Some will recognize that the systems of equations that will have infinite solutions means that both equations must be equivalent. Some may consider the graph of two parabolas and recognize that if the parabolas have a common vertex but are not the same everywhere, then the system has one solution. As students consider systems of quadratic equations that have two solutions, expect them to think about translations and/or reflections, such as $y = x^2$ and $y = -x^2 + 2$.

4. **Annie graphed a system of linear equations and concluded that there is no solution. Do you agree? Why or why not?**

(See Reproducible 6, Annie's Graph.)

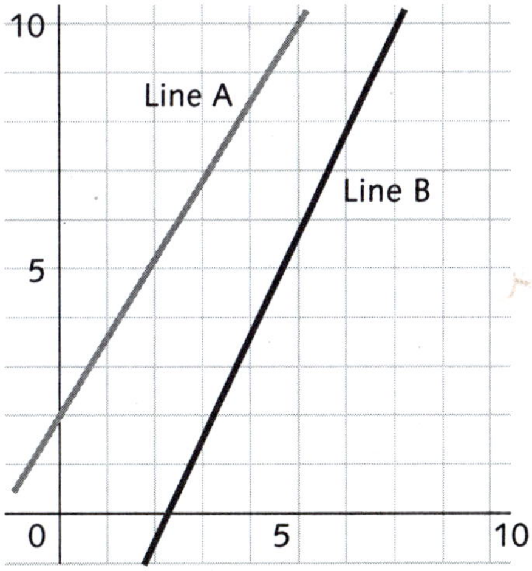

This question raises a couple of issues for students to grapple with. One is that for two graphed lines in a system to have no solution, the lines must be parallel. Therefore, students should demonstrate how they know (beyond "eyeing it") that these two lines do not have the same slope. In addition, a system of equations takes into account solutions that exist within the entire set of solutions for each equation, which extend beyond the bounds of any graph.

5. What are two different ways to generate a one-variable equation from the following system using substitution?

$$3y = 2x - 5$$

$$5(2x - 5) - 3y = 4$$

This question raises the prospect that substitution does not require an equation to be in terms of a single variable (e.g., solved for y). Upon close inspection, students should notice that the expression $2x - 5$ is in both equations, allowing $3y$ to be substituted in the bottom equation for $2x - 5$ (resulting in $5(3y) - 3y = 4$). In addition, $3y$ is present in both equations, enabling $2x - 5$ to be substituted for $3y$ in the bottom equation (resulting in $5(2x - 5) - (2x - 5) = 4$). This question presents an opportunity for students to compare and contrast these two approaches.

6. Filipe proposes that a system with two quadratic equations can have three solutions. Do you agree or disagree? Explain your answer.

This question can spark a debate. Some students will interpret this question as having *exactly* three solutions, and they should be encouraged to explain why that is impossible. However, other students can assert that since the quadratic equations can be equivalent, they can have infinite solutions, which means that three solutions are possible.

7. Charlotte wants to come up with three two-variable equations so that when all of the potential systems of two equations are created, one has no solution, one has one solution, and one has infinite solutions. Is this possible? Why or why not?

This is not possible. If a system with two of the equations has infinite solutions, then the two equations are equivalent. That means that when they are both paired with the third equation, the same solution will result—even if the two equivalent equations do not "look the same." This question encourages students to rethink what infinite solutions mean and is benefited by considering its graphical representation.

8. What is a possible second equation of the following system so that the system has no solution? Exactly one solution? Exactly two solutions? Infinite solutions?

$$y = 8x - 3$$

$$\underline{\quad ? \quad}$$

This question encourages students to consider how other equations would interact with a given linear function. Since exactly two solutions are not possible with a system of linear equations, then students need to consider other potential equations that will allow for only two solutions, such as absolute value equations or quadratic equations.

9. Consider the functions $f(x)$ and $g(x)$ graphed here. How could either function be transformed (through translation or reflection) so that the system of equations has exactly one solution? Two solutions? For each case, write the transformed system of equations and verify the number of solutions.

(See Reproducible 7, Graph of Transformed Functions.)

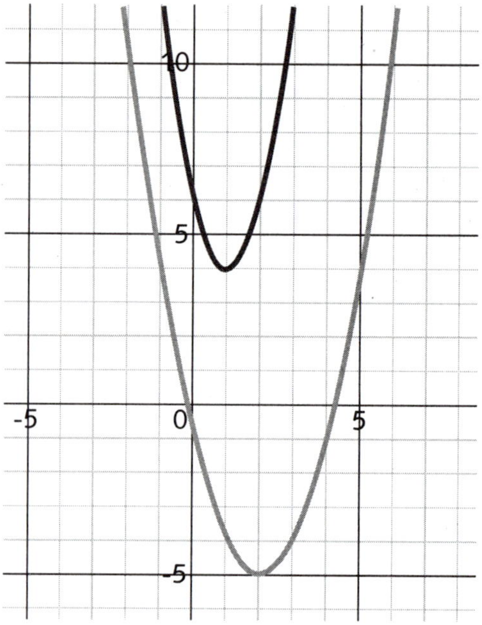

$f(x) = 2x^2 - 4x + 6$

$g(x) = x^2 - 4x - 1$

This question enables students to use a variety of combinations of transformations to accomplish this task. For example, students could create a system with one solution by transforming $g(x) \to g(x + 1) + 9$ or by $g(x) \to g(x) + 7$ (which can be validated by solving the system).

10. Andre argues that he can look at this system of quadratic equations and know a possible solution. What do you think he is doing and why does it work? Generate another system of quadratic equations that enables a solution to be read from the equations.

$$y = x(x + 3) - 2$$

$$y = (x - 1)(x + 3) - 2$$

The point (-3, -2) must make both equations true. This is because $(x + 3)$ is a factor of the first term of both expressions, leaving $y = -2$.

11. How is Equation A below the same or different than the system of Equations B? Explain.

A	B
$3(2y - 10) + 4y = 7$	$x = 2y - 10$ $3x + 4y = 7$

Expect students to see that Equation A is the result of substituting $2y - 10$ for x in the system of equations in B. This may cause some students to think A and B are the same. However, Equation A has only a solution for y, whereas the system of Equations B has a coordinate (x, y) solution.

12. Miriam says that $y = 4$ and $x = 2$ is a system of perpendicular lines that intersect at (2, 4). What is another system of perpendicular lines with the same solution? Explain how you found your answer.

Students can approach this question in a variety of ways. One way is to find another equation that passes through (2, 4), such as $y = x + 2$, and then finding the line perpendicular to that line through (2, 4). A generalization, using point-slope form, could be $y = a(x - 2) + 4$ and $y = -\frac{1}{a}(x - 2) + 4$.

Rewriting Expressions

Experiences at This Level Will Help Students To

- use an area model to connect different expressions
- connect the distributive property $a(b + c) = ab + ac$ to processes for multiplying binomials

Reproducibles are available in a downloadable, printable format. See page xx for directions about how to access them.

Materials

- Area of Rectangle (Reproducible 8)
- Expressions Venn Diagram (Reproducible 9)

Good Questions and Teacher Notes (pages 74–77)

1. **What are different expressions that represent the area of the rectangle with dimensions as shown here? Find at least three.**

(See Reproducible 8, Area of Rectangle.)

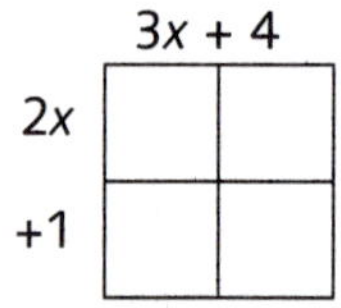

The area model enables students to recognize why and how each term of both factors must multiply each of the terms of the other factor. Asking students to write at least three expressions encourages students to write the area both in terms of a product (i.e., $(2x + 1)(3x + 4)$) and as a sum (e.g., $6x^2 + 8x + 3x + 4$ and $6x^2 + 11x + 4$).

2. **What are different ways you can rewrite the expression $3xy^2 - 12x^3$? Generate at least three different ways to rewrite the expression.**

This question offers a lot of different ways to factor including factoring just 3 or $3x$ or recognizing that $(y + 2x)$ and $(y - 2x)$ are also factors.

3. Use a rectangle to show that $a(b + c)$ always equals $ab + ac$. How can you extend the diagram to show that $a(b + c + d) = ab + ac + ad$? What other identity must be true?

This question moves students to generalize the distributive property, which is often unnecessarily limited to a monomial multiplied by a binomial. Some students may extend the diagram to $a(b + c + d + e)$ while others may extend it to be $(a + e)(b + c + d)$. Encourage different diagrams and expressions to be shared and analyzed.

4. Two students are having a debate. One student says that $(2 - 3x)(x + 5)$ can be rewritten as $2(x + 5) - 3x(x + 5)$ and another says it cannot. Which student is correct and why?

This question connects multiplying binomials, which some students see as a disconnected process without basis, and the distributive property. For sense making, you may want to ask students how they can justify that these expressions are equivalent with a geometric diagram.

5. Explain why $4x^2 - 12x + 5 = 4x^2 - 12x + 5 + 4 - 4$. How can this way of rewriting help to show that $4x^2 - 12x + 5 = (2x - 3)^2 - 4$?

This question encourages students to recognize that rewriting quadratic equations strategically enables them to create a special quadratic (in this case, a perfect square trinomial).

6. Are these statements sometimes, always, or never true? Explain.

$4(\frac{x}{4} + 2) = x + 8$ $\quad$ $4(\frac{x}{4} + 2) = x + 10$ $\quad$ $4(\frac{x}{4} + 2) = 2x + 8$

This question provides an opportunity for students to see that when an expression is rewritten as an equivalent expression, they are always equal. Equations for which the two sides are not equivalent are either never or sometimes true.

7. Katya expanded the expression $(2x + 5)^2$ and got $4x^2 + 25$. Do you agree or disagree? Explain using an area model.

This question draws attention to a common error where students believe they can square both terms.

8. What expressions can go in each part of this Venn Diagram? Explain.

(See Reproducible 9, Expressions Venn Diagram.)

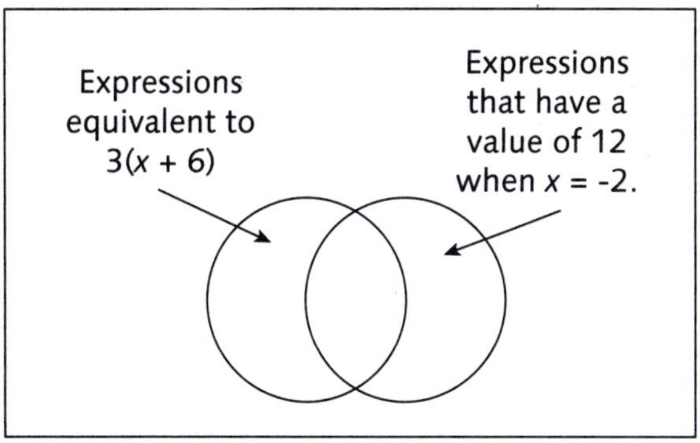

Students may be surprised that no matter how they rewrite $3(x + 6)$ to become equivalent, the expression will have a value of 12 when $x = -2$. Therefore, all expressions in the left circle belong in the overlapping region. However, there exists some expressions that are not equivalent to $3(x + 6)$ that have the same value. Thus, although equivalent expressions must have the same value for a given x-value, those that have the same value are not necessarily equivalent.

Solving Inequalities

Experiences at This Level Will Help Students To

- understand that a solution is a set of values that make an inequality true
- represent solutions to inequalities graphically as a solid region (in the case of two dimensions) or a connected set of points (in the case of one dimension), unless there is no solution
- understand that solution regions are typically bounded by solutions to equalities
- understand that inequalities can be strictly inequality or can include equality
- identify inequalities that have no solutions

Reproducibles are available in a downloadable, printable format. See page xx for directions about how to access them.

Materials

- Solutions Venn Diagram (Reproducible 10)
- Gerald's Graph (Reproducible 11)
- Inequalities Venn Diagram (Reproducible 12)

Good Questions and Teacher Notes (pages 79–86)

1. **Name at least one solution (x, y) of this inequality for each quadrant, if possible. For example, is there a solution in Quadrant I (for which $x > 0$ and $y > 0$)?**

   Quadrant II (for which $x < 0$ and $y > 0$)?

   Quadrant III (for which $x < 0$ and $y < 0$)?

   Quadrant IV (for which $x > 0$ and $y < 0$)?

$$x - 2y < 6$$

This question focuses students' thinking on the nature of solutions to two-dimensional inequalities. Students should recognize why there are infinite solutions to inequalities (because there are infinite points to select in each of these quadrants that make the inequality true).

2. **Write at least two inequalities that have solutions $x = 3$, $x = -1$, $x = -2.5$. Write at least one inequality that does not have these solutions.**

This question focuses students' attention on the fact that an inequality has a set of values that make it true (even if that set is empty) and that it must be designed to make the entire solution set true. So students should ask themselves, "What inequalities would automatically make all of these true?" Solutions can be varied. Although many students may use simple linear inequalities such as $x > -3$ or $x < 5$, other more complex inequalities will also span these solutions, such as $|x - 1| < 4$ or $(x - 3)(x + 2.5) \leq 0$.

3. What set of solutions would describe each region of this Venn Diagram? What would go in the region outside the two circles? Explain how you know.

(See Reproducible 10, Solutions Venn Diagram.)

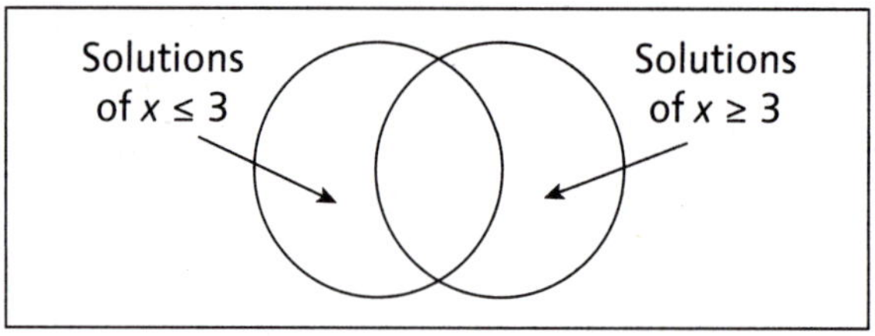

This question encourages students to compare and contrast the solutions of $x \le 3$ and $x \ge 3$ to recognize that the only solution in common is $x = 3$ and that this would go in the intersection of the Venn diagram. Then, because the combined solution set spans all reals, no real number could go outside the two circles. However, if your students have learned of imaginary numbers, then they should recognize that these would go outside the circles.

4. Katie says that both (-3, 4) and (5, 6) are solutions to the same inequality of the form $y > mx + b$. What other points must also be solutions? How do you know?

This question requires students to use what they know about the solution region of a strictly linear inequality (namely, that it is bounded by a linear equality and consists of all points on one side of the line). If both (-3, 4) and (5, 6) are solutions, then all the points in between these two points must necessarily be solutions as well (because otherwise, these two points would be on opposite sides of the boundary line). Therefore, students can name any point on the line connecting (-3, 4) and (5, 6) between the two points. In addition to these points, students may notice that there are infinitely more solutions possible by increasing y-values of the given solutions. That is, since $4 > -3m + b$, then $5 > -3m + b$ as well, which means that (-3, 5) is also a solution (along with all points (-3, k) where $k > 4$). Other examples of this phenomenon include (5, 7) and (1, 5.01).

5. How are the following the same? How are they different? Explain.

$$x(x-2) > 0 \qquad x(x-2) < 0$$

This question forces students to recognize that the same boundaries (in this case, the boundary $y = x(x - 2)$) can lead to different solution sets. In the first inequality, the solutions are the points that are $x < 0$ and $x > 2$, while in the second inequality, the solutions are the points $0 < x < 2$.

6. In testing the point (0, 0), Rashad found that it made an inequality true. Is this enough to know the entire solution region? Why or why not?

Answers to this question depend on the level of students' knowledge of inequalities and their possible regions. Students should recognize that if (0, 0) does not lie on a boundary of the solution region, then they would at least be able to identify a region of the solution set. However, if (0, 0) lies on the boundary, then it is not possible to identify which side of the boundary is included in the solution to the inequality.

7. Write an inequality that has the same solution as the following inequality.

$$x^2 < 9$$

This question can help students relate absolute value with quadratic inequalities. Since the solution set is the set of points $-3 < x < 3$, another way to write this set is with the inequality $|x| < 3$. Thus, when solving quadratic inequalities of the form $x^2 < a^2$, students can rewrite this as $|x| < a$.

8. Gerald says that he can "see" several inequalities looking at the following graph. What inequalities might he see?

(See Reproducible 11, Gerald's Graph.)

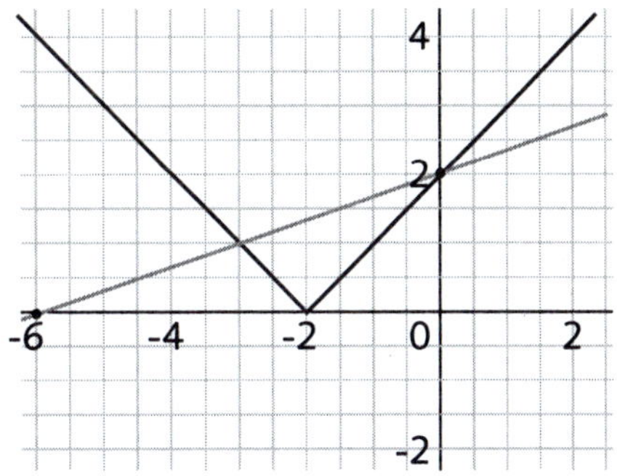

Possible inequalities are:

$$|x + 2| \le \frac{1}{3}x + 2$$

$$|x + 2| \ge \frac{1}{3}x + 2$$

$$|x + 2| \ge 0$$

This question requires students to connect the solution set of an inequality with the graph of a system. Students should connect the inequality $|x + 2| \le \frac{1}{3}x + 2$ with the question "Where in the graph is the absolute value function less than (or below) the linear function?" This interpretation should enable them to see why the solution to $|x + 2| \le \frac{1}{3}x + 2$ must be $-3 < x < 0$. Students should also recognize that graphs can help them understand why some inequalities can have a solution of all reals (such as in the case of $|x + 2| \ge 0$).

9. Draw a number line that represents the set of points in each region (*a*, *b*, *c*, and *d*) of this Venn Diagram.

(See Reproducible 12, Inequalities Venn Diagram.)

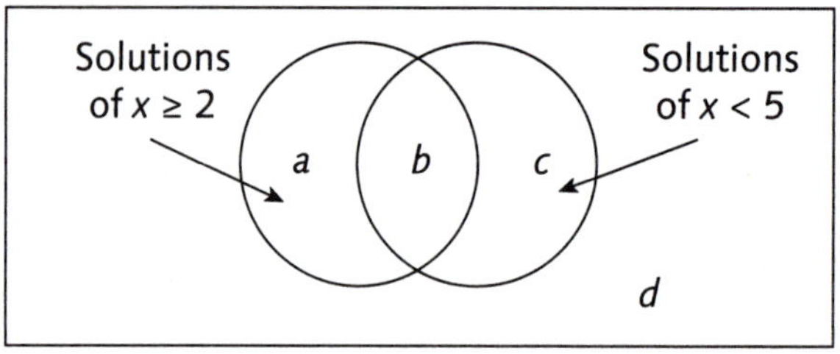

This question requires students to recognize all the possible combinations of solutions generated from these two inequalities and lays a foundation for systems of inequalities. For example, (b) contains the set of points $2 \leq x < 5$ and would be represented by the following number line.

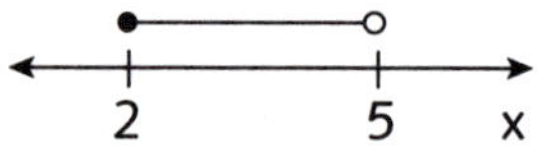

Then (a) must be the set of all points greater than 2 that are not represented in the set for (b), leading to the following number line.

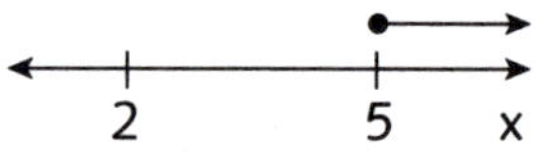

Similarly, (c) must be the set of all points less than 5 that are not represented in the set for (b), leading to the following number line.

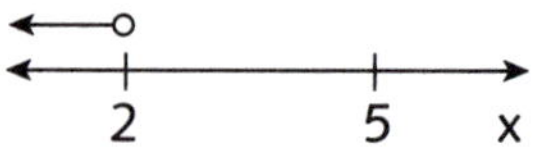

By elimination, all real solutions are now accounted for. So the solution set for (d) must be no solutions, represented by the following number line.

10. Is it possible for a linear inequality to have no solution? If it is possible, then explain why. If not possible, then give an example.

This question will likely generate discussion. Students need to consider what they know about linear inequalities. Since all linear inequalities have a boundary that is a linear function, and since they all include either one or both of the regions of the plane that does not include the line (both sides of a linear function would be included in a solution in the case of $y \neq mx + b$), then it is not possible to have no solution when the lines do not have the same slope. However, when the lines are parallel, it is possible to have no solution, such as when $y < x - 5$ and $y > x + 3$.

11. What inequalities could go in each category in the following table? For each, generate at least one inequality that contains a variable x.

Never True	Sometimes True	Always True

This question enables students to compare the qualities of inequalities that have different types of solutions (none, some, all). Encourage students to be creative. Inequalities that are always true include those such as $x^2 \geq 0$. Inequalities that are never true can be generated with absolute values or quadratics, but even simple inequalities, such as $x > x$, are never true.

CHAPTER 6

GOOD QUESTIONS FOR FUNCTIONS

Students often enter high school with a limited concept of functions (Conney, Beckmann, Lloyd 2010). For example, they may think that all functions are linear or quadratic. Or, they may think of functions only in terms of algebraic expressions or equations. As they progress through high school, it is important that students develop a more robust understanding of functions as they explore "classic" families of functions (i.e., linear, quadratic, exponential, rational, trigonometric, and polynomial) and others as well. Questions in math class should aim to help students understand functions as descriptions of how related quantities vary together, identify the key characteristics of particular families of functions, connect different representations of the same function (e.g., a table to a graph), and combine and transform functions.

Linear Functions

Experiences at This Level Will Help Students To

- write a linear equation in standard and slope-intercept form
- use linear equations to model relationships
- connect different representations of linear functions
- identify a function as linear or nonlinear

Reproducibles are available in a **download-able, print-able format.** See page xx for directions about how to access them.

Materials

- AB's Graph (Reproducible 13)
- Mystery Values and Graph (Reproducible 14)

Good Questions and Teacher Notes (pages 89–99)

1. **The graph of a linear equation passes through quadrants I, III, and IV only and contains the point (6, -1). What might be its equation?**

After verifying the solutions, ask students to generalize about all answers. Do students notice that they all have a positive slope and a negative *y*-intercept? You can modify this question by specifying the form students should use.

2. **Write a standard form equation for a linear relationship whose graph has a negative slope and a positive *y*-intercept.**

One way to answer this equation is to write a slope-intercept form equation with the given constraints and then convert it to standard form.

Post students' answers and ask students what all of the answers have in common. Students may be surprised to note that the values of A, B, and C (for $Ax + By = C$) all have the same sign. Spend some time talking about why this is true—particularly why it makes sense that positive values of A, B, and C correspond to a graph with a *negative* slope.

3. AB's teacher asked him to find the standard form equation for the linear relationship graphed here.

AB wrote "$5x + 6y = \frac{6}{5}$." Do you agree or disagree with AB's answer? Explain.

(See Reproducible 13, AB's Graph.)

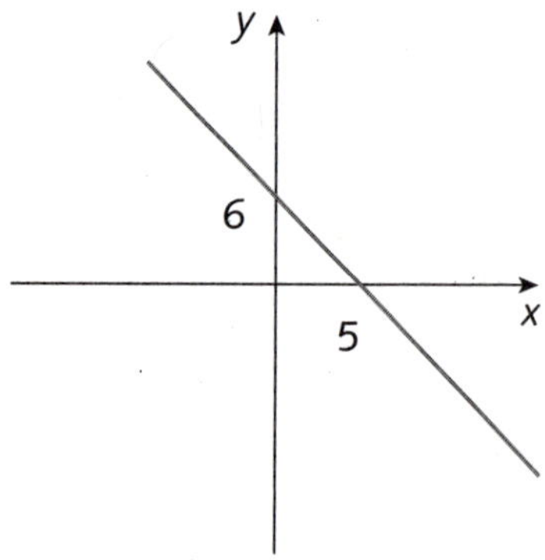

AB's answer reveals a common incorrect response with linear equations in standard form. Many students think that the x-intercept and y-intercept are equivalent to the values of A and B, respectively, in the standard form equation. One way students can detect this error is to find the x- and y-intercepts of AB's equation and compare those values with those given on the graph. For example, if students substitute y with 0 and solve for x, they will see that $5x = \frac{6}{5}$; $x = \frac{6}{25}$ whereas the x-intercept on the graph is 5. Or, introduce intercept-intercept form ($\frac{x}{5} + \frac{y}{6} = 1$) to reinforce that the other coordinate is always 0.

4. Replace C in the linear equation $4x - 3y = C$ so that the y-intercept of the graph of this equation is an integer.

Post students' answers and ask students what all of the answers have in common. Students should note that C is always a multiple of 3.

5. BC drove a total of 120 miles in a little less than 2.5 hours. What might have been her average speed in miles per hour?

One way to answer this question is to divide 120 by 2.5 to find the average speed 48 miles per hour. Since BC drove 120 miles in less than 2.5 hours, she must have been traveling slightly faster than 48 miles per hour. One possible average speed is 50 miles per hour (e.g., at this rate she would travel 120 miles in 2.4 hours).

6. Your school wants to raise $3,000 by selling tickets to a play. They will charge students slightly less than adults. How much might they charge for each type of ticket?

This problem context lends itself well to a standard form linear equation. Once students derive their answers, ask them to explain what each term and each variable means in the context of the problem. Specifically, it is important that students can differentiate whether a variable expression represents an amount of money or a number of tickets. For example, one possible answer is $5x + 7y = 3{,}000$ where x is the number of student tickets sold, $5x$ is the income from the student tickets sold, y is the number of adult tickets sold, and $7y$ is the income from the adult tickets.

7. CJ borrowed $500 from his parents to pay for a new bicycle. He will repay his parents with a monthly payment plan (the same amount of money each month). He wants to be done paying back his parents in less than two years. How much might he pay each month? Write an equation that could be used to figure out his remaining balance after *x* number of months.

One possible answer is $y = 500 - 50x$, where y is the money still owed, x is the number of months, $50 is the amount he pays back each month, and $500 is the cost of the bike. Extend this question by asking students to determine how long it will take CJ to pay back the money for the monthly rate they set in their equation.

8. A high school graduation ceremony will include three speeches. Then, diplomas will be awarded to 200 graduates—names will be read one at a time in alphabetical order. Write an equation that could be used to estimate how long after the start of the ceremony a particular graduate's name will be called. (Adapted from www.illustrativemathematics.org.)

This is an open-ended problem that will require students to make several assumptions. For example, students will need to estimate how long the three speeches will take as well as how long it takes to read each name (and whether or not they should include time for applause). Students will also need to think carefully about their units of time as well as how to define their variables.

One possible answer is as follows: Assume the three speeches take one hour. Assume one name is read every 10 seconds: $y = \frac{1}{6}x + 60$, where x is the ordinal position of the graduate in alphabetical order and y is the time (in minutes) from the start of the ceremony.

9. A biathlete finished a biathlon — a two-mile swim and a half-marathon — in approximately three hours. Sketch a graph that could show her total distance over time.

Students may need to look up typical rates for swimmers and runners before they answer this question. Students will also need to make a decision about whether they want to show the athlete moving at a constant rate of change.

10. A first-place finisher of a 40-meter dash crossed the finish line three seconds before the second-place finisher. Assuming each runner ran at a virtually constant speed, sketch a graph that could be used to model each runner's distance over time.

Once students make their sketches, ask them to describe what they notice. (One possible graph is given below.) For example, the line showing the speed of the second-place finisher is less steep than that of the winner.

One way to extend this question is to ask students to calculate each runner's average speed in meters per second for the graphs they sketched.

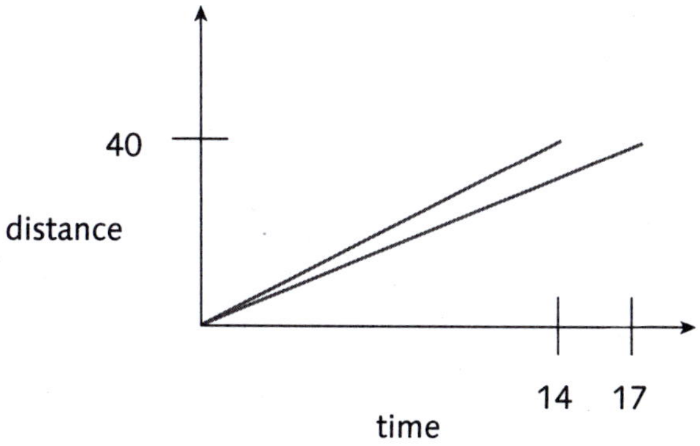

11. Complete this table of values so that it shows data from a linear relationship.

x	y
-4	
-1	
3	
7	

One way to answer this question is to create a linear equation, substitute x for the given value, and solve for y. Another possible strategy is to fill in two values for y, use those to calculate the rate of change, and then fill in the remaining two rows accordingly.

You can extend this question by asking students to make observations about the linear relationship they created. For example, is the linear relationship also a direct variation? If they graphed their line, is the slope positive or negative?

12. Write a table of (x, y) values for a linear equation whose graph has a negative slope and a positive y-intercept.

Do students consider including a row where $x = 0$ as a way to find the y-intercept of the line?

13. Complete this table of values to create a linear equation with a slope equal to $\frac{2}{3}$. Find the y-intercept for the equation you created.

x	y

One possible answer is as follows. The y-intercept for this equation is 0.

x	y
-9	-6
-3	-2
0	0
3	2

14. Is it possible that this graph and table of values represent the same linear equation? Why or why not?

(See Reproducible 14, Mystery Values and Graph.)

x	y
-4	-2
-1	-5
5	-11
8	-14

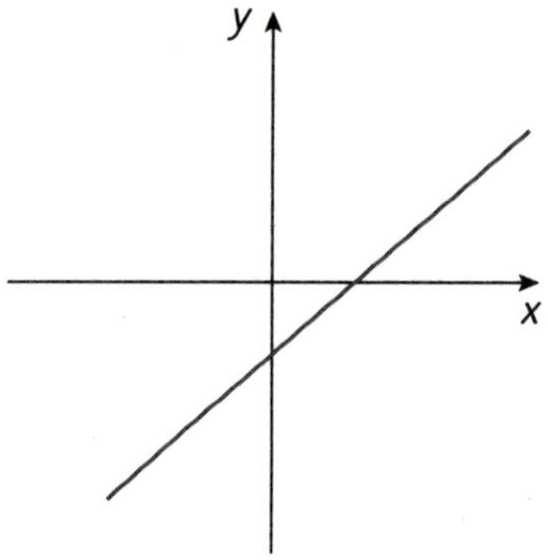

Students can use a variety of strategies to determine that no, it is not possible that the graph and table of values represent the same linear equation. For example, the table shows data from a linear relationship with a negative rate of change or slope but the slope of the line on the graph is positive.

15. Complete this table of values so that it shows a linear relationship whose graph has a y-intercept of -4.

x	y
-2	
0	
1	
7	

One way to answer this question is to begin by filling in the second row to show the y-intercept (0, -4). Then, students might choose any value for the row above or below and use that information to determine the rate of change for their linear relationship and fill in the other blanks accordingly.

16. DC looked at the following table of values. He said, "This table cannot show a linear relationship because the rate of change is not constant." Why might DC say this? What is wrong with his reasoning?

x	y
-8	-1
-4	1
0	3
8	7

It seems that DC looked to see if the table represented a direct variation, which is a type of linear relationship. DC's reasoning isn't wrong, but it is inconclusive.

17. **Complete the table of values on the left so that it shows data from a linear relationship and the table of values on the right so that it shows data from a nonlinear relationship.**

Linear Relationship

x	y
-4	
-1	
5	
8	

Nonlinear Relationship

x	y
-4	
-1	
5	
8	

This question focuses students' attention on the importance of rate of change when determining if a relationship is linear or nonlinear. Specifically, a relationship between two variables is linear if there is a constant rate of change between the variables. You might extend this question to motivate a discussion about relation versus function. For example, you could ask, "Consider the values you chose for the table on the right; do they show data from a relation or a function?" Or, "Complete the table of values on the right so that it shows data from a relation that is not a function."

18. Examine the geometric formulas that follow. Choose two to compare and contrast.

Circumference of a Circle	$C = \pi d$
Area of a Circle	$A = \pi r^2$
Perimeter of a Square	$P = 4s$
Area of a Square	$A = s^2$
Volume of a Sphere	$V = \frac{4}{3}\pi r^3$

Be sure students identify the independent and dependent variables in each formula and recognize that π is a constant—and not a variable. Encourage students to consider the formulas as members of families of functions.

Quadratic Functions

Experiences at This Level Will Help Students To

- write a quadratic equation in standard, factored, and vertex form
- convert between forms of a quadratic equation
- connect the form of an equation of a quadratic function to characteristics of the related parabola
- use the form of a quadratic equation to make predictions about its solutions
- compare properties of two quadratic functions using their equations and graphs
- solve a quadratic equation via factoring, completing the square, and using the quadratic formula

Reproducibles are available in a downloadable, printable format. See page xx for directions about how to access them.

Materials

- Graph of $f(x) = x^2$ and Mystery $h(x)$ (Reproducible 15)
- Graph of Mystery Quadratic Function (Reproducible 16)
- Graph of Translation of $g(x) = 2x^2$ (Reproducible 17)

Good Questions and Teacher Notes (pages 101–107)

1. Kevin wrote the expression $4s^2 - 36$ to represent the area of a rectangle. What might be the dimensions of this rectangle?

Do students recognize this expression as the difference of squares? For example, the expression can be rewritten as $(2s - 6)(2s + 6)$ which can be used to represent the side lengths of the rectangle. Still, other students might point out that the dimensions could be 4 and $s^2 - 9$. Other possible answers include $(4s - 12)(s + 3)$ or $(4s + 12)(s - 3)$.

Do students see that the rectangle exists only for values of $s > 3$?

2. Write at least two quadratic equations in the form $ax^2 + bx + c = 0$—where a, b, and c are nonzero integers—that has two solutions, $x = \frac{2}{5}$ and $x = \frac{3}{4}$.

One possible answer is $(5x - 2)(4x - 3) = 20x^2 - 23x + 6 = 0$. However, this quadratic equation could also be multiplied by any scalar (except 0) and the solutions would remain the same.

As students work on this question, check to see if they first write the equation in factored form and then apply the distributive property to rewrite the equation in standard form.

3. Tuhina looked at a quadratic equation in vertex form and thought to herself, "Based on the graph of the related function, I know the equation has no real solutions." What might be the equation Tuhina was looking at?

Possible answers include: $0 = (x - 2)^2 + 3$, $0 = -0.25(x + 4)^2 - 1$, and $0 = 17(x - 5)^2 + 2$. Any quadratic equation of the form $f(x) = a(x - p)^2 + q$ where a and q are both positive or both negative will have no real solutions because their graphs never touch the x-axis.

4. A quadratic equation with integer solutions is of the form $ax^2 + bx + c = 0$ and begins with the expression $x^2 + 18x$. Find a possible integer value for c.

Possible answers include: 17, 32, and 45. Post several answers on the board. After verifying each solution, ask the students to describe what they all have in common. Students should note that for all nonzero values of c, c has two factors whose sum is equal to the value of b.

5. A quadratic function has an equation that can be expressed as the difference of squares but its parabolic graph is *not* symmetric about the y-axis. Write an equation that satisfies this description. Find the roots of the equation you wrote.

Post several answers on the board. After verifying each solution, ask the students to describe what they all have in common. All solutions should be of the form $(x - p)^2 - q$, where q is a square number, for example, $(x - 4)^2 - 9 = 0$.

6. A quadratic equation is of the form $ax^2 + bx + c = 0$. Find integer values for a, b, and c such that the minimum value of the function is (-1, 3).

Once students find a possible equation, ask them to write their equation, the equation of the related function, and a sketch of the graph of that function on the board. Ask students what they notice about all of the graphs. Can they use the graphs to generalize about the number of real solutions to each of the equations?

7. Write an equation—in factored form—of a quadratic function with a line of symmetry $x = -3$.

Since the roots of a quadratic equation are symmetric about its line of symmetry, we can use the midpoint between the roots to find the line of symmetry. For example, the equation $(x + 5)(x + 1) = 0$ has a line of symmetry at $x = -3$.

8. Write an equation of a quadratic function that has two integer roots each less than 0.

As an extension, ask students to sketch the graph of the quadratic equation they created.

9. A problem in TL's math book asked her to find the maximum point on the graph of a quadratic function. TL completed the square to find that the maximum point was (4, 12). What might have been the equation of the function as it appeared in TL's math book?

Since the parabola has a maximum point, we know a (in $a(x - p)^2 + q$) has a negative value. Since its maximum point is (4, 12), we also know the values of p and q. One possible solution is $-(x - 4)^2 + 12 = -x^2 + 8x - 4 = f(x)$.

10. Find a solution for c such that the equation $x^2 + 14x + c = 0$ has no real number solutions.

One possible way to answer this question is to use the quadratic formula and create a discriminant that is negative. For example, we could write $\sqrt{b^2 - 4ac} = \sqrt{14^2 - 4c} = \sqrt{196 - 4c}$ and choose any value of $c > 49$.

11. Find a quadratic equation in the form $ax^2 + bx + c = 0$ (where a, b, and c are nonzero integers) that has two irrational solutions.

This is a good question to assess students' understanding of the difference between irrational numbers and imaginary numbers.

12. Joan looked at a quadratic equation in the form $ax^2 + bx + c = 0$ and said, "I can tell just by looking at this that one solution is negative and one is positive." What equation might Joan have been looking at?

All answers should be of the form $ax^2 + bx + c = 0$, where c is a negative integer, assuming that a is positive.

Post several answers on the board. After verifying each solution, ask the students to describe what they all have in common. Students should note that all equations have a negative value for c.

13. This graph shows $f(x) = x^2$ and the graph of another quadratic function, $h(x)$. What might be the equation of the function $h(x)$?

(See Reproducible 15, Graph of $f(x) = x^2$ and Mystery $h(x)$.)

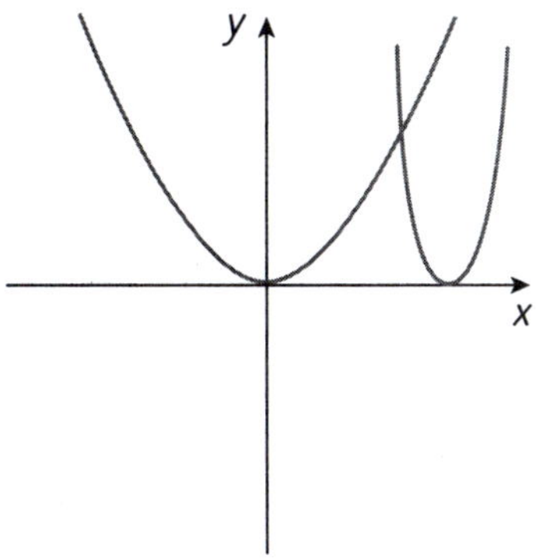

Students can use translations to answer this question. If they note that $h(x)$ is narrower, that tells them that a (in the term ax^2) is greater than 1. One possible answer is $h(x) = 4(x - 3)^2$.

14. Examine the following sketch of the graph of a quadratic function. What might be the equation of this graph?

(See Reproducible 16, Graph of Mystery Quadratic Function.)

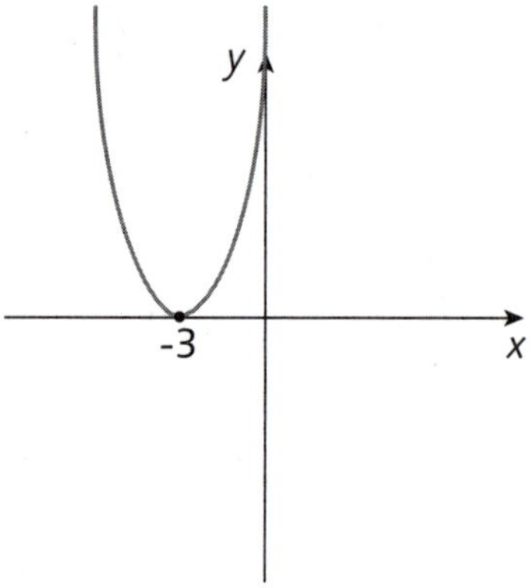

One possible answer is $f(x) = 2(x + 3)^2$.

As an extension, ask students to find any roots of the equation they created.

15. **The following graph is a translation of the graph $g(x) = 2x^2$. What might be its equation?**

(See Reproducible 17, Graph of Translation of $g(x) = 2x^2$.)

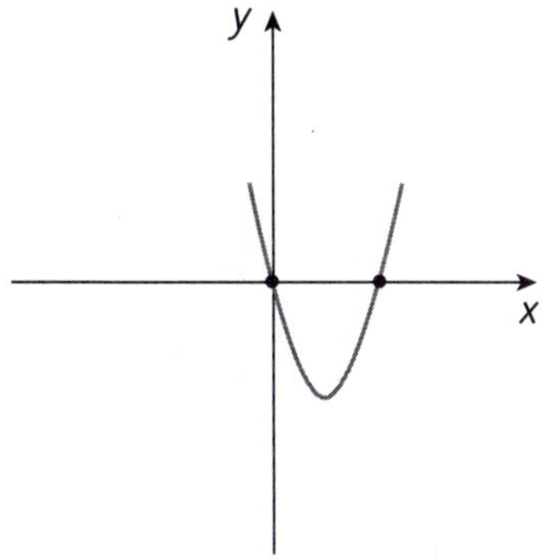

One possible answer is $f(x) = 2(x - 3)^2 - 4$.

As an extension, ask students to find the roots of the equation they created.

Exponential and Logarithmic Functions

Experiences at This Level Will Help Students To

- compare and contrast linear functions and exponential functions
- identify growth/decay factor and initial value of an exponential function
- represent geometric sequences using exponential equations
- for exponential models, express as a logarithm the solution to $ab^{ct} = d$, where a, c, and d are numbers and the base b is 2, 10, or e; evaluate the logarithm using technology
- understand and compare the properties of classes of exponential and logarithmic functions

Reproducibles are available in a downloadable, printable format. See page xx for directions about how to access them.

Materials

- GC's Graph (Reproducible 18)
- Graph of Mystery Exponential Function (Reproducible 19)
- Graphs of Two Mystery Exponential Functions (Reproducible 20)
- Graphs of Two Mystery Functions (Reproducible 21)
- Graphs of Three Mystery Functions (Reproducible 22)
- $y = \log_B(ab^x)$ Mystery Graphs (Reproducible 23)
- Mystery Graphs of $y = f(x)$ and $y = 2f(x)$ (Reproducible 24)

Good Questions and Teacher Notes (pages 109–123)

1. GC looked at this sketch of the exponential equation and said, "I know for sure that the growth factor is 4."

Do you agree or disagree with GC's statement? Explain.

(See Reproducible 18**, GC's Graph.)**

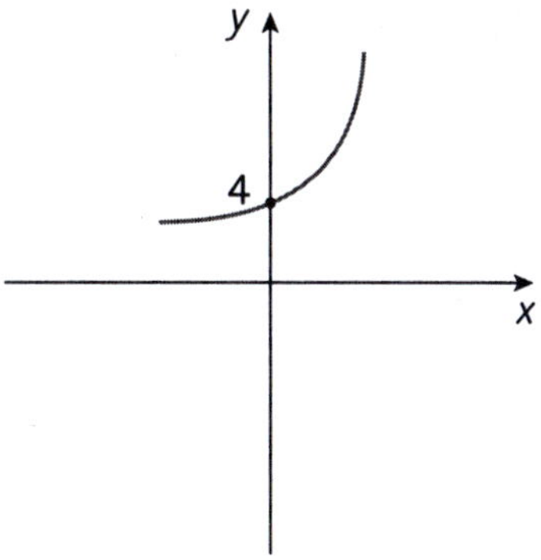

This question targets the common incorrect response that the y-intercept of the graph of an exponential function reveals the value of b in the equation $y = ab^x$.

Once students recognize this mistake, they may make another mistake and conclude that the sketch must represent the function $f(x) = 4(1^x)$. So be on the lookout for this mistake as well.

2. The graph of a certain exponential function is shown here. Create a table of (x, y) values that could represent this function.

(See Reproducible 19, Graph of Mystery Exponential Function.)

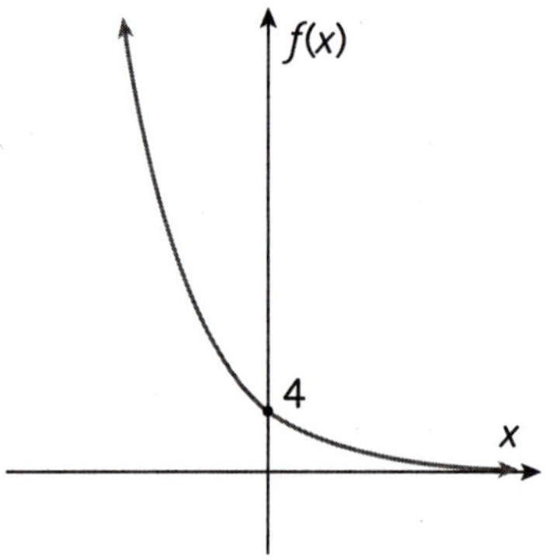

This question encourages students to use characteristics of the graph to reason about the constraints of the exponential relationship. For example, students should note that the y-intercept of the graph reveals that the ordered pair (0, 4) is included in this relationship. In addition, since the graph shows an exponential decay, the values of y will decrease (by a constant factor) as x increases.

3. The graphs of two exponential functions are shown here. Write two equations that could model these graphs.

(See Reproducible 20, Graphs of Two Mystery Exponential Functions.)

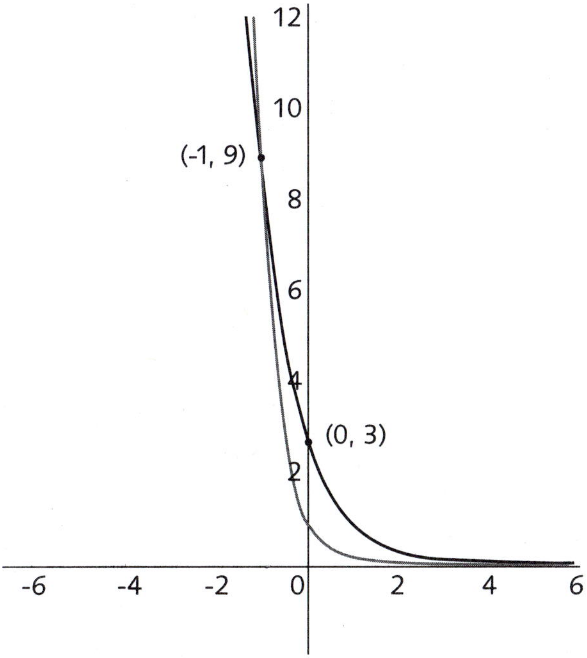

Students are given two points for one graph, which allows them to derive the equation $y = 3(\frac{1}{3})^x$ for the black graph. There are many options for the grey graph but any equation should show a slower rate of change and include the solution (-1, 9) (e.g., $y = (\frac{1}{9})^x$).

4. TZ looked at an exponential equation and said, "I can tell just by looking at the equation that one solution is (1, 90)." What might be the exponential equation?

Students can derive a possible equation by reflecting on the general form of an exponential function: $f(x) = ab^x$.

One possible solution is $f(x) = 45(2^x)$.

As an optional extension, ask students to write the general form of any exponential function whose graph contains the point (1, 90). (Answer: $f(x) = \frac{90}{b}b^x$.)

5. DA's teacher asked her the following question: The graph of an exponential function goes through (1, 6) and (2, 12). What is the equation of this function?

DA asked, "How can the function be exponential? I thought two points always define a line?"

If you were DA's math partner, how might you respond?

If the graph of a function is a line, two points are enough information to derive the equation of that function. However, if we are given two points on the graph of a function, we cannot conclude that the function is linear. Infinitely many functions could go through the same two points—including an exponential function.

Follow-up questions include: Why can the equation for an exponential relationship be derived from only two points? Which type(s) of functions cannot be derived with only two points?

6. Two functions were graphed on the same coordinate grid as shown here. What might be the equation of each graph?

(See Reproducible 21, Graphs of Two Mystery Functions.)

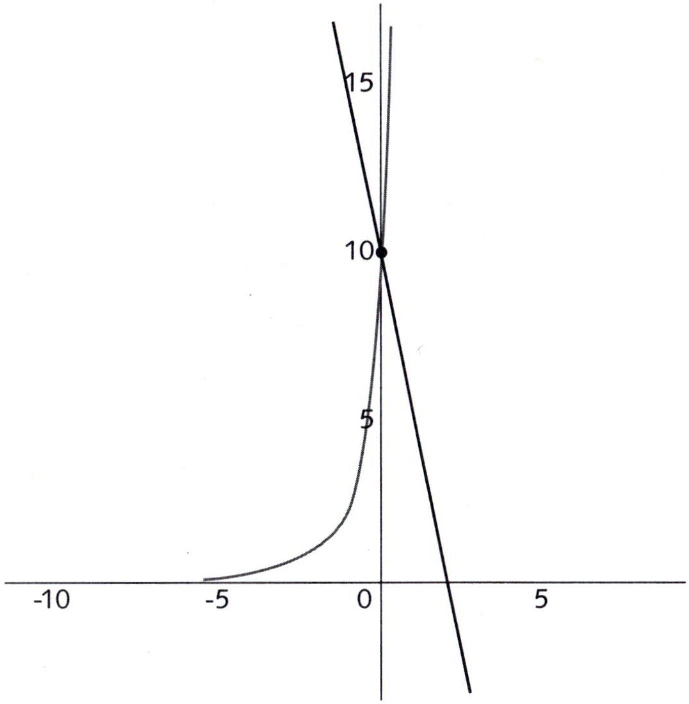

One possible answer is as follows: $y = 10 - 5x$ and $y = 10(5^x)$

Students may make the mistake of writing $y = 10(1^x)$ for the equation of the nonlinear graph. But by definition, the base of an exponential is greater than 0 and not equal to 1. One way to help students see this error (or others) is to check their answers using graphing technology.

7. JK's teacher asked her the following question: Compare and contrast $f(x) = 12(2^x)$ and $f(x) = 2(12^x)$.

JK said, "These represent the same exponential function because they both can be simplified to $f(x) = 24^x$." Do you agree or disagree with JK's answer? Why?

Push students to refute JK's answer beyond referencing rules of exponents. Encourage students to use the terms *initial value* and *growth factor* when explaining how the equations are similar and different. For students who still are confused, have them evaluate both functions for $x = 2$, so they can see that $12 \times 2 \times 2$ will not be equal to $2 \times 12 \times 12$. Also, be sure that students observe that one point of intersection proves nothing.

8. An exponential growth function and an exponential decay function share the point (-1, 2). What might be the equation of each function?

Students might find it helpful to sketch a graph of the functions described in the problem.

You may want to ask students to post their equations on the board. Then, as a class, discuss how they are similar and different. Encourage students to use the terms *initial value* and *growth factor* in their explanations.

9. A certain bacterium multiplies by a certain factor every hour such that, six hours after the first bacterium appears, there are approximately 2 million bacteria in the colony. Write a function to model this situation.

Students who have not yet studied logarithms can explore this question using the general form of an exponential equation, $y = ab^x$, and the given information in the problem. Specifically, if $2{,}000{,}000 = b^6$, then $b \approx 11.22$ and the function can be modeled by the equation $y = 11.22^x$.

In Questions 10 to 19, the term "logarithmic function" refers to a function in the form $y = f(x) = \log_b x$ unless otherwise noted.

10. Abe was wondering if the function $y = \log_{(-2)} x$ is valid. Since $(-2)^3 = -8$, he figured that $y = \log_{(-2)}(-8) = 3$. Bea was wondering if the function $y = \log_1 x$ is valid. Since $1^x = 1$, she figured that $y = \log_1 1 = 1$. What do you think about their reasoning?

Students can explore the discontinuous nature of logarithms with negative bases and when the base is 1 to understand why the potential values for the base are restricted, that is, $b > 0$, $b \neq 1$.

11. Three functions are graphed on the following axes. What might be the equations for each function?

(See Reproducible 22, Graphs of Three Mystery Functions.)

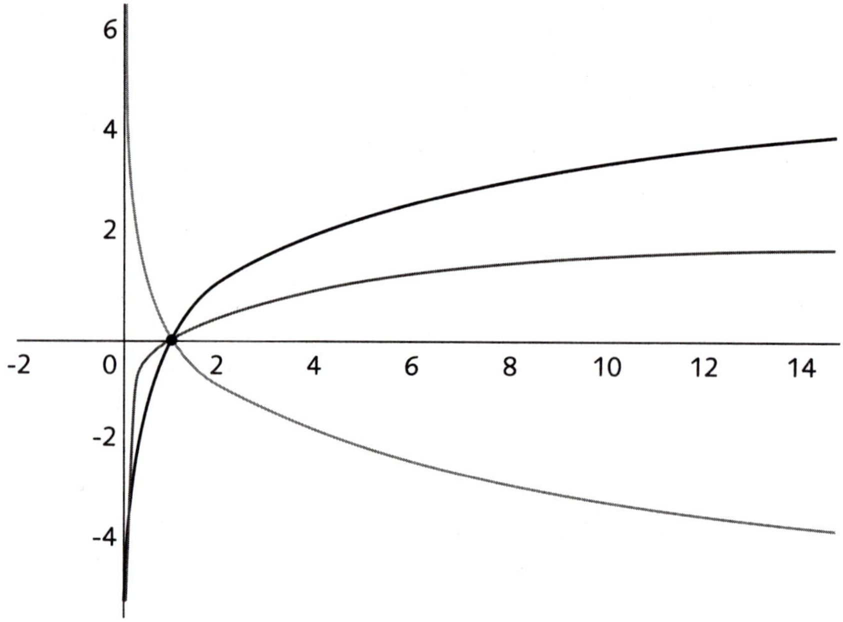

One possible set of answers is:

$y = \log_2 x$

$y = \log_5 x$

$y = \log_{\frac{1}{2}} x$

Students can discuss the common features that the x-intercept is always 1, and that the domain is $x > 0$. Also, they should notice the symmetry over the x-axis of $y = \log_2 x$ and $y = \log_{\frac{1}{2}} x$.

12. Abe, Bea, and Cia were talking about taking the logarithm of the terms of a particular geometric sequence.

What might the original sequence have been?

How could they have gotten such 'different' results?

- Abe got the sequence 6, 8, 10, 12, 14, . . .
- Bea got the sequence 3, 4, 5, 6, 7, 8, . . .
- Cia got the sequence -1.5, -2, -2.5, -3, -3.5, . . .

Students should realize that the logarithm of a geometric sequence is an arithmetic sequence and the common difference depends on the base of the logarithm and the common ratio of the geometric sequence. This will help students understand the inverse structure.

13. For each of the following equations, write as many possible solutions (a, b) as you can:

- $\log_b a = 2$
- $\log_b a = \frac{3}{2}$
- $\log_b a = -3$
- $\log_b a = 0$

Students need to see the structure in order to find multiple solutions. For example, students might recognize that (4, 2) and (9, 3) are both solutions to $\log_b a = 2$. Then students should discuss the rationale they used in creating their solutions and any patterns they notice across all the solutions. For example, for $\log_b a = 2$, a will always be the square of b.

14. Create an example to demonstrate that the logarithm of the outputs of an exponential function $f(x) = ab^x$ will result in a linear function.

Does the base of the logarithm need to be the same as the exponential function?

What are the slope and y-intercept?

There are several insights that students can gain from understanding this characteristic. First, it introduces the effect of inverse functions. Second, you may want to point out how this offers a strategy to adjust exponential data so that they become linear in order to allow for easier regression.

15. The following graph shows four functions of the form $y = \log_B(ab^x)$, where a, b, and B are not necessarily unique.

What might be the equation for each graph?

(See Reproducible 23, $y = \log_B(ab^x)$ Mystery Graphs.)

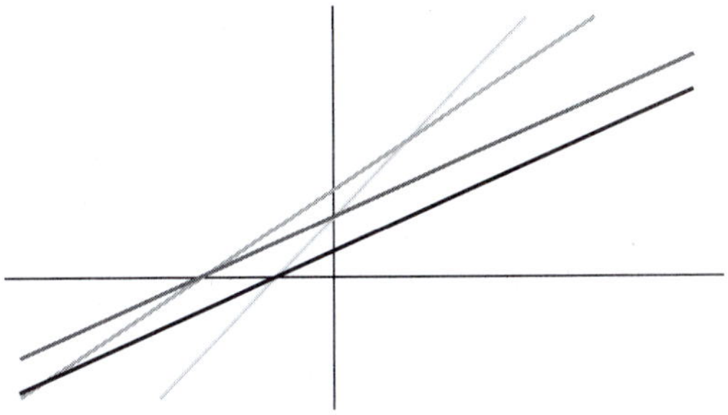

One possible set of answers would be:

$y = \log_2(2 \cdot 2^x)$

$y = \log_5(5 \cdot 2^x)$

$y = \log_5(2 \cdot 2^x)$

$y = \log_3(5 \cdot 2^x)$

Students can discuss why the log of an exponential function is linear as well as the relationship between the values of a, b, and B.

16. In the following graph, the graphs of $y = f(x)$ and $y = 2f(x)$ are shown. What might be the equation for each function's graph? Can you write the equations as two logarithmic functions; that is, not a logarithmic function and a dilated logarithmic function?

(See Reproducible 24, Mystery Graphs of $y = f(x)$ and $y = 2f(x)$.)

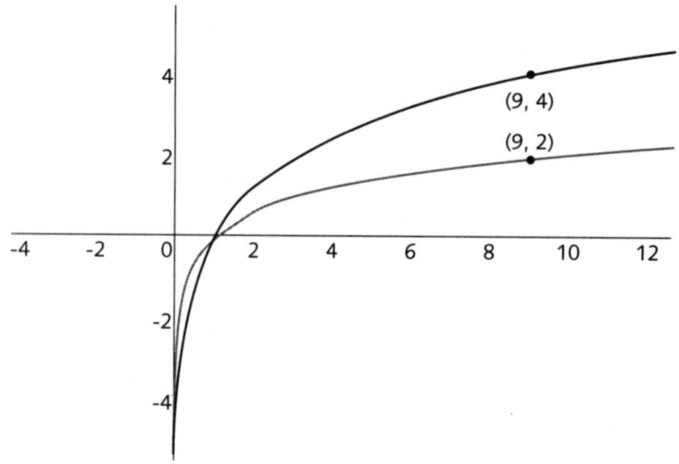

Possible solutions are:

$y = \log_3 x$

$y = 2\log_3 x$

$y = \log_{\sqrt{3}} x$

Students often struggle to grasp the interplay between the coefficient of a log expression and the base of the logarithm. Students may also suggest that $y = \log_3 x^2$ is a solution. It is a valid function, but is not itself a logarithmic function. Instead, it is a composition of a log and quadratic function, which can be another direction for the discussion.

17. A forensic scientist wrote the following equation: $100e^{0.07t} = 200$.

What might Abe, Bea, and Cia be thinking?

- Abe says that he can solve the equation using trial and error.
- Bea says that she can solve for t algebraically.
- Cia has some thoughts as to what the equation might represent.

A possible solution: Abe might be thinking that since $100e^{0.07t} \approx 187.8$ for $t = 9$ and $100e^{0.07t} \approx 201.4$ for $t = 10$, that the solution is $t \approx 9.9$.

Bea could solve it in two main ways: $t = \frac{\log 2}{0.07 \times \log e}$ if she was using the properties of logs, or she might write $t = \frac{\ln(2)}{0.07}$ if she was re-expressing the equation using natural logarithms.

Finally, Cia could be thinking that the scientist had a concentration of 100 percent of the virus that was growing continuously at a rate of 7 percent per hour and was curious as to when the virus would double (in about 9.902 hours). Or the scientist was looking at what she owes (in thousands) if her student loans of $100,000 compounded continuously at 7 percent interest over 10 years without payment.

18. Since $4 = 2^2 = 8^{\frac{2}{3}}$, Abe, Bea, and Cia were wondering if 4 could be written as an exponential expression with any base between 0 and 10 (inclusive). Abe said he could do it for all but two of the integers between 0 and 10, inclusive. Bea said that her expression had a negative exponent.

What expressions might Abe and Bea have looked at to write 4 using bases between 0 and 10?

Cia noticed that all of the other bases required irrational exponents. Do you agree?

Students might see that Abe realized that the base could *not* be either 0 or 1, but that other solutions can exist in the form of $\log_b 4 = \frac{\log 4}{\log b}$, for b = 3, 5, 6, 7, . . . 10. So, for example, $4 = 3^{\log_3 4}$ is a way to express 4 using 3 as a base. Bea, on the other hand, noticed that bases between 0 and 1 yield negative exponents, which reinforces the form of the graph of a logarithmic function. Cia's point is true for the integer bases, but an infinite number of possibilities exist for bases between 0 and 1. So, for example, $4 = \frac{1}{2}^{-2}$ is a way to express 4 using $\frac{1}{2}$ as a base which requires a negative exponent.

19. Abe, Bea, Cia, and Dee are each given a set of two points in the coordinate plane and asked to write equations that could pass through the two points. Abe says that he can write a linear and an exponential function, but that he cannot find a logarithmic function containing the two points. Bea reports that she can find a linear and a logarithmic function, but not an exponential. Cia reports that she can find all three types of functions. Dee says that she can only find a linear function. What set of points might each have been given, and what functions did each write?

This question helps students apply the concepts of domain and range for the different functions. You can expand the discussion to include other types of functions, too.

Note:
Abe could have had at least one point with a negative x-coordinate.

Bea could have had at least one point with a negative y-coordinate.

Cia could have had two points in the first quadrant.

Dee could have had two points in the third quadrant.

A what-if discussion of other possibilities may be productive here.

Rational Functions

Experiences at This Level Will Help Students To

- describe a rational function in terms of its key characteristics (e.g., [removable] discontinuity, x-intercept, y-intercept, vertical asymptotes, horizontal asymptote, and limit)
- understand that rational functions may contain discontinuities that are asymptotic or removeable (holes)
- recognize rational functions in different forms (e.g., $R(x) = \frac{N(x)}{D(x)} = Q(x) + \frac{R(x)}{D(x)}$) and understand that each form contributes insight as to the nature of the function

Good Questions and Teacher Notes (pages 125–130)

1. Write the equation for a rational function whose domain includes all real numbers except 3.

> You can adapt this question by changing the constraints so that the domain excludes exactly two values. You might also consider asking students to write the domain using interval notation.

2. ZZ looked at a rational number function and said, "I can tell the domain is all real numbers." His math partner LL said, "How can that be? Rational functions always have at least one value excluded from their domain." Which student do you agree with? Why?

> ZZ is correct; although rational functions often contain discontinuities, they don't always. For example, the function $y = \frac{x}{x^2 + 5}$ is a rational function whose domain is all real numbers (i.e., no number makes the denominator equal to 0).

3. Write an equation for a rational function that doesn't have a y-intercept.

> After posting and verifying a variety of answers, ask students to generalize. For example, students should note that for all equations, the denominator equals 0 when 0 is substituted for x (e.g., $y = \frac{x+6}{x}$). This also means that 0 is excluded from the domain of each function.

4. Some of the numbers from the rational function here were accidentally erased. What might be the missing values so that the function is discontinuous at exactly two points?

$$y = \frac{x}{x^2 + ?x + ?}$$

One way to answer this question is to create a quadratic expression that can be factored. For example, $y = \frac{x}{x^2 + 8x + 15} = \frac{x}{(x + 3)(x + 5)}$, which is discontinuous at $x = -3$ and $x = -5$.

5. Write the equation for a rational function with x-intercepts (0, 0) and (4, 0).

One possible answer is $y = \frac{x^2 - 4x}{x^2 + 5}$.

6. Some of the numbers from the rational function following were accidentally erased. Do you still have enough information to determine the x-intercepts?

$$y = \frac{x(x + 3)}{x^2 + ?x + ?}$$

No, we do not. The x-intercepts of a function have coordinates $(a, 0)$. For rational functions, we can find the x-intercepts by setting the numerator equal to 0 because $0 \div n = 0$ (where $n \neq 0$). But this is only true if the fraction is in simplest form. In the example given, if the denominator is $y = \frac{x(x + 3)}{x^2 + 4x + 3}$ the fraction simplifies to $y = \frac{x}{x + 1}$, which has one x-intercept at (0, 0) and one "hole" at $(-3, \frac{3}{2})$. But if the equation is $\frac{x(x + 3)}{x^2 + 5x + 3}$ the fraction is in simplest form and has x-intercepts at (-3, 0) and (0, 0).

7. A rational function has a vertical asymptote at $x = 2$ and another at $x = 1$. What might be its equation?

A vertical asymptote marks a value of x for which the function is not defined. Students can find a possible answer by creating a rational function whose denominator is equal to 0 (but not its numerator) when $x = 2$ or 1. For example, $y = \frac{2x^2 - x + 5}{x^2 - 3x + 2}$.

8. Fill in the blanks to create true statements about rational functions.

a. All rational functions ____________________.

b. Only some rational functions ____________________.

c. If a rational function ____________________, then ____________________.

You may need to provide additional scaffolding for this question by encouraging students to include the following terms in their answers: (removable) discontinuity, x-intercept, y-intercept, vertical asymptotes, horizontal asymptote, limit (e.g., if a rational number has a vertical asymptote, then it has a (nonremovable) discontinuity). Note: A removable discontinuity is synonymous with a hole.

9. Alex, Britt, and Caleb write out rational functions such that Alex's function has a hole at (3, 4); Britt's has a vertical asymptote that passes through (3, 4); and Caleb's has a horizontal asymptote that passes through (3, 4). What might their functions be? How are they similar? How are they different? Is it possible that any two of the functions are the same?

A possible solution could be that:

Abe wrote the function $f(x) = \dfrac{(x+1)(x-3)}{x-3}$,

Britt wrote the function $g(x) = \dfrac{(x+1)(x-2)}{x-3}$, and

Caleb wrote the function $h(x) = \dfrac{4(x+1)}{x-3}$.

This question asks students to think about which structures lead to removable discontinuities (i.e., holes), which lead to vertical asymptotes, and which lead to horizontal asymptotes.

10. Near and Far are looking at the graph of the same rational function on graphing software. Near is looking at a zoomed-in graph and states that the function has one vertical asymptote. Far is looking at a zoomed-out graph and says that the function looks like a line with a negative slope. What rational function could they both be viewing?

This question focuses on the end behavior and the local behavior. Students should notice that the numerator must have a degree one larger than the denominator. So, for example, if they were looking at a graph of $y = (1 - x)(\frac{1+x}{x})$ it would show a vertical asymptote at $x = 0$, but when zoomed out would look like the line $y = -x$.

11. **The graph of a rational function has vertical asymptotes at $x = -1$ and $x = 6$. What might this graph look like? Create a sketch.**

You might also have students write a possible equation for the graph they sketch. One possible equation and graph are:

$$y = \frac{x - 1}{x^2 - 5x - 6}$$

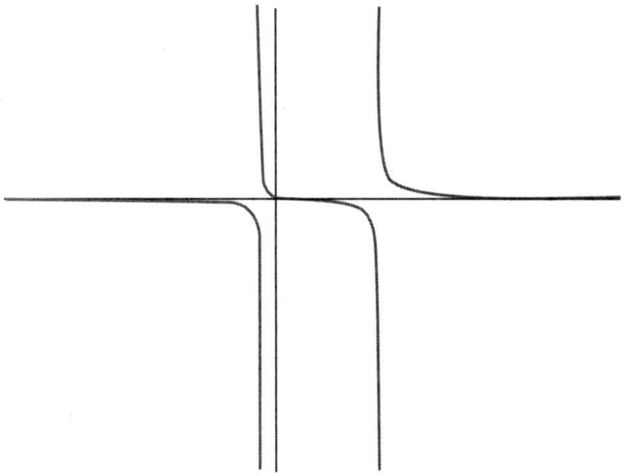

12. Doana and her partner were working on a homework set on rational functions. Doana wrote the answer to one problem as follows:

$$y = \frac{x}{x+1},\ x \neq -3$$

Her partner, Leah, said, "I think you made a mistake. The domain can include -3 because -3 plus 1 doesn't equal 0. We only exclude values for *x* that make the denominator equal to 0." Doana said, "I checked my answer in the back of the book and it's correct."

What problem might Doana have been working on?

Doana might have been working on a problem that asked her to remove a discontinuity and rewrite the function rule. For example, $y = \frac{x(x+3)}{x^2+4x+3} = \frac{x}{x+1}$, $x \neq -3$.

13. The limit of a rational function, *f*, as *x* approaches 3, is 2. Write a possible equation for this function.

To find the limit of a function, *f*, at the *x*-value *a*, we replace all values of *x* with *a* and simplify. So, one way to answer this question is to create a rational function where the output or *y*-value is 2 when $x = 3$. One possible answer is $y = \frac{x+5}{x+1}$.

Polynomial Functions

Experiences at This Level Will Help Students To

- use basic elements (i.e., coefficients, roots, degree) of a polynomial function to predict features of its graph (i.e., number of intercepts, curves, increasing or decreasing, end behaviors)
- understand that when a polynomial function is rewritten as a product, this equivalent expression can reveal new properties of the graph of the function (i.e., that it passes through a certain point or has a certain x-intercept)
- recognize that there are multiple ways to rewrite each polynomial that contains a product of polynomials if you don't require the remainder to be 0
- relate different polynomials via transformations

Reproducibles are available in a downloadable, printable format. See page xx for directions about how to access them.

Materials

- Representations of Polynomial $P(x)$ and $Q(x)$ (Reproducible 25)
- Relationships between a Cubic and Linear Function (Reproducible 26)
- Polynomial $P(x)$ Area Model (Reproducible 27)
- Selina's Graph and Table (Reproducible 28)
- Graph of $g(x) = 2x^3 + 9x^2 + 3x - 4$ (Reproducible 29)

Good Questions and Teacher Notes (pages 132–141)

1. If possible, sketch a graph of a cubic polynomial that has no x-intercepts. If not possible, explain why not.

This is not possible. This question requires students to grapple with what they know about cubic polynomials (for example, how many turns can they have?) to figure out whether this cubic can exist. A natural follow-up question is to have students describe the conditions under which polynomials can have no x-intercepts.

2. Find at least two different polynomials that have the following features:

- It has at least three zeros, two of which are equal
- It is symmetric with respect to the y-axis
- As x increases to infinity, y decreases to negative infinity

This question has many answers. However, from the given information, students could place the double root on the line of symmetry (i.e., the y-axis) or place it at some other x value where $x \neq 0$. If they choose to put the double root on the line of symmetry, then possible solutions are of the form $P(x) = ax^2(x - b)(x + b)$ where $a < 0$, where additional pairs of roots are equidistant from the origin. If they choose to put the double root at $x \neq 0$, then possible solutions are of the form $P(x) = a(x - b)^2(x + b)^2$ again where $a < 0$. Note that the degree must be an even number to have symmetry with respect to the y-axis.

3. Compare the polynomials represented here. How are they the same or different? Explain.

(See Reproducible 25, Representations of Polynomial $P(x)$ and $Q(x)$.)

$P(x) = 2x^2 - 13x + 18$ $Q(x) =$

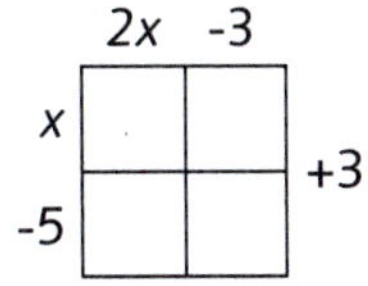

Students should recognize that $Q(x)$ is $(2x - 3)(x - 5) + 3$ and that this simplifies to $P(x)$. Therefore, $Q(x)$ and $P(x)$ are equivalent polynomials and have the same graph. They are different in how they are written. The different forms of the expressions reveal different aspects of the graph of this polynomial. When written as $P(x)$, it can be seen that the graph has a y-intercept of (0, 18). When written as $Q(x)$, it can be determined that the graph passes through the points (1.5, 3) and (5, 3) by finding the zeroes of the binomial product.

4. After being asked to compare $f(x) = x^2$ and $g(x) = \frac{1}{2}x^2$, Madison and Milton got into an argument. Madison thinks that $g(x)$ is *wider* than $f(x)$, while Milton says it is *shorter*. Who is correct and why? Or, if both are correct, how can you understand how each is looking at the relationship of f and g?

This question is an opportunity to connect visual information in the graph and the way students describe the shape of functions with the symbolic expression of the functions. When comparing the graphs of f and g, many students will refer to g as wider. Yet symbolically, $g(x)= \frac{1}{2}f(x)$ and therefore every y-value of $g(x)$ is half the value of the corresponding y value of $f(x)$. Students should describe how "shrinking" the polynomial vertically by multiplying by $\frac{1}{2}$ has the same result as stretching the graph horizontally.

5. After dividing $P(x) = 3x^2 + 5x - 8$ by $(x + 2)$, Brad figured out that $P(x)$ can be rewritten as $(x + 2)(3x - 1) - 6$. What does this new form of the expression tell him about the graph of $P(x)$? Explain.

The goal of this question is for students to recognize that the new form of the expression reveals new information about $P(x)$, for example that the graph of $P(x)$ is that of $(x + 2)(3x - 1)$ after being shifted down six units, so $P(x)$ must pass through the points (-2, -6) and ($\frac{1}{3}$, -6).

6. **The graph shows the polynomial $Q(x) = (x - 1)(x - 2)(2x + 3) - 4$ and the line $y = -4$. What is the relationship of the way $Q(x)$ is written and the graph? Explain your ideas.**

(See Reproducible 26, Relationships between a Cubic and Linear Function.)

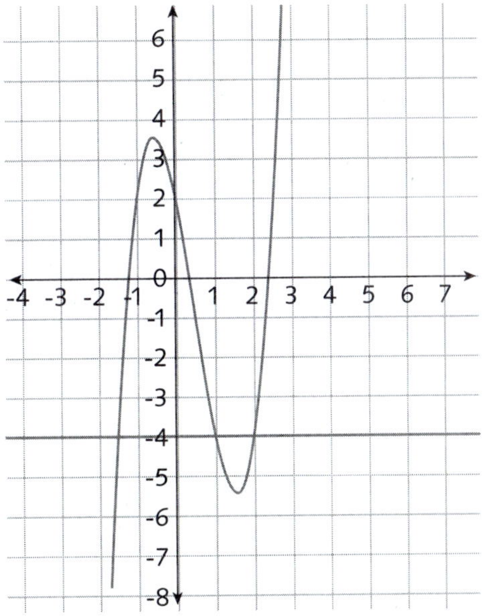

The goal of this question is for students to orient their thinking about a polynomial expression as a potential intersection of two functions. In this case, $Q(x)$ is the sum of a cubic $f(x) = (x - 1)(x - 2)(2x + 3)$ and a constant function $g(x) = -4$. Therefore, you may want to ask students how the factored product in $Q(x)$ is connected with the graph, by asking "What happens at $x = 1$? $x = 2$? and so on until students recognize that the solutions of the factored terms reveal the x-values of the intersection of $Q(x)$ and $y = -4$.

7. The graph of $Q(x) = (x - 1)(x - 2)(2x + 3) - 4$ is shown here. How can you rewrite the expression of $Q(x)$ so that it is instead in the form $Q(x) = (\quad)(\quad)(\quad) + 2$? What information exists in the graph to support you?

(See Reproducible 26, Relationships between a Cubic and Linear Function.)

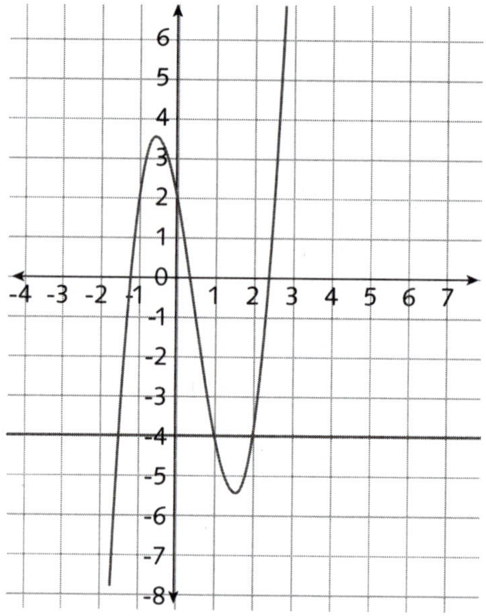

This question starts to get at the ideas of the remainder theorem. That is, when a polynomial $Q(x)$ is divided by an expression in the form $(x - a)$ for some real value a, then the remainder of that division is the value of $Q(a)$. The reason for this is that when $Q(x)$ is rewritten as $Q(x) = (x - a)(\ldots) + r$, then $Q(a) = (a - a)(\ldots) + r = 0 + r = r$. Since the graph of $Q(x)$ passes through the points (-1, 2) and (0, 2), then students can determine that $Q(x) = (x + 1)(x)(\ldots) + 2$. Then by dividing $Q(x) - 2$ by $x(x + 1)$, they can complete the expression: $Q(x) = (x + 1)(x)(2x - 5) + 2$.

8. If $P(x) = x(x - 3)(2x + 3) - 5$, name at least three points on $P(x)$. Explain how you can know from the way in which the expression is written.

Students may attempt this problem by starting with x-values such as $x = 0$ and substituting them in to get the corresponding y-value. However, if they use $x = 0$ or $x = 3$, they may notice how the outputs of this function easily become -5. The purpose of this question is to start students thinking about how writing a polynomial as a partial product supports the recognition that when $x = 0$, $x = 3$, and $x = -\frac{3}{2}$, then $P(x) = -5$. Thus, three of the points on $P(x)$ that can be discerned from the expression are (0, -5), (3, -5), and $(-\frac{3}{2}, -5)$. Seeing a polynomial this way supports student reasoning around the remainder theorem.

9. Cora figured out that a certain polynomial $Q(x)$ could be written in two different ways: $Q(x) = (x - 1)(3x + 8) - 4$ and $Q(x) = (x - 2)(3x + 11) + 10$. She then concluded that $Q(x)$ must have a zero between 1 and 2. Is she correct? How could she know without looking at a graph?

Since the remainder theorem guarantees that the polynomial passes through the points (1, -4) and (2, 10), and because polynomials are continuous functions, then there must be an x-intercept between $x = 1$ and $x = 2$. This means that the polynomial must have a zero between 1 and 2. This offers an opportunity to introduce the Intermediate Value Theorem.

10. The polynomial $P(x)$ can be represented with the following area model, or as the expression $(x + 1)(5x - 2) - 4$. Find at least two other ways to express $P(x)$.

(See Reproducible 27, Polynomial $P(x)$ Area Model.)

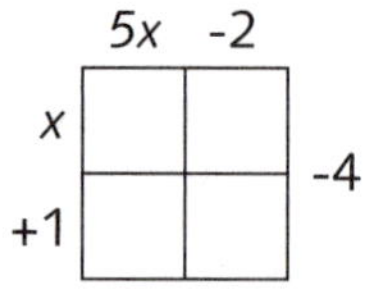

This question has multiple answers. However, one way many students may answer it is to expand the expression to find that $P(x) = 5x^2 + 3x - 6$. However, students can also pick different dimensions for the area model to find a different way to write $P(x)$ as a partial product. For example, $P(x)$ also can be written as $(x - 2)(5x + 13) + 20$ or as $x(5x + 3) - 6$. You may want to have a discussion where students share out all the different forms to build an organized list from which students can draw additional conclusions about $P(x)$.

11. Write the expression of a quartic polynomial that passes through the points (-3,2), (-1,2), and (1,2). Is there more than one possible quartic polynomial that passes through these points? Explain how you know.

There are an infinite number of solutions to this question. However, given that all of these given points have $y = 2$, then the quartic must be able to be written as $Q(x) = (x + 3)(x + 1)(x - 1)(\ldots) + 2$. Therefore, students can replace the missing term in $Q(x)$ with any binomial of the form $ax + b$ to find a quartic that passes through the given points. This question requires students to reverse their thinking with the remainder theorem. Rather than use x values to find y values when given a polynomial, students can use related coordinates to build a polynomial.

12. Selina started to build a table of expressions for the graph of $P(x)$. How can you complete her table? And how is her table related to the graph of $P(x)$?

(See Reproducible 28, Selina's Graph and Table.)

$P(x)$
$(x - 1)(x + 2) - 1$
$(x - 0)(x + 1) - 3$
$(x - 2)(x + 3) + 3$
(?)(?) – ?
(?)(?) – ?

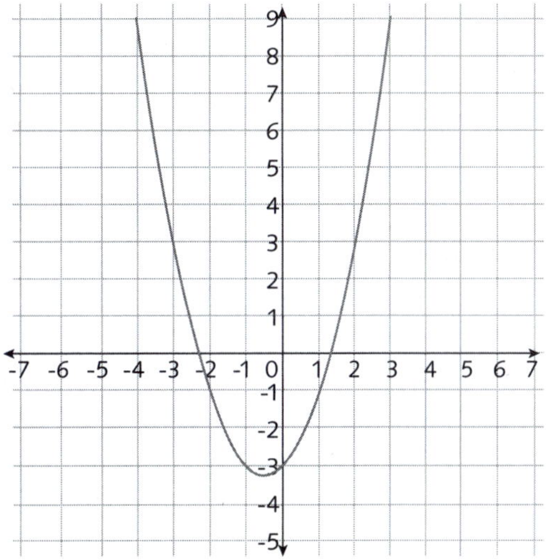

This question helps students see that each graph of a function can be represented by an infinite number of "rules" or equations when taking into consideration different vertical shifts. Give students time to find patterns between the table and graph. Students should notice that for $y = 9$, the function $P(x)$ passes through lattice points, offering the expression $(x - 3)(x + 4) + 9$ as another way of writing $P(x)$. They may want to complete the table with $P(x) = (x - 4)(x + 5) + k$ for some k value, but this was intentionally left off the graph. The students can determine k by figuring out $P(4)$ or $P(-5)$ for any of the other expressions. Consider asking follow-up questions that press for reasoning on why the pattern works.

13. Eddie took the line $f(x) = x$ and translated it down 3 units. Estelle instead translated the line $f(x) = x$ to the right 3 units. How are their resulting functions the same or different? Explain.

The transformations coincide and both can be represented as $f'(x) = x - 3$. However, when translating to the right, this can be connected to the function $f'(x) = (x - 3) + 0$ and the translation down 3 units can be connected to the function $f'(x) = (x) - 3$. These resulting transformations represent different mappings but collectively involve the same locus (or group) of points. In a discussion of this question, you may want to ask "Where does the point (0, 0) move to in Eddie's transformation? In Estelle's? What does that tell us about the relationship of these transformations?"

14. Given the graph of $g(x) = 2x^3 + 9x^2 + 3x - 4$ here, sketch the graph of $h(x) = 2(x+1)^3 + 9(x+1)^2 + 3(x+1) - 4$ and $k(x) = 2x^3 + 9x^2 + 3x + 1$.

(See Reproducible 29, Graph of $g(x) = 2x^3 + 9x^2 + 3x - 4$.)

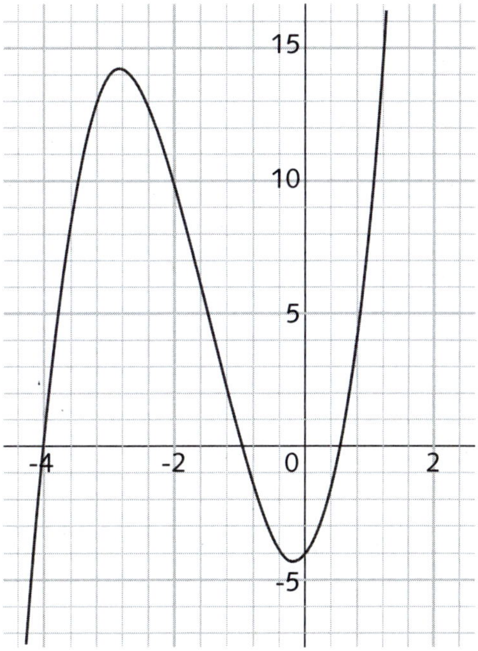

This question draws attention to the structure of polynomials and how these are related by translations. When the function $g(x)$ is translated to the left 1 unit, the result is $h(x)$. When the function $g(x)$ is translated up 5 units, the result is $k(x)$. You may want to extend this discussion to ask students to come up with their own variation of $g(x)$ that is the result of a translation to challenge the class to graph.

Trigonometric Functions

Experiences at This Level Will Help Students To

- understand radian measure as an arc length on a unit circle
- use special triangles to determine, geometrically, the sine, cosine, and tangent for $\frac{\pi}{3}$, $\frac{\pi}{4}$, and $\frac{\pi}{6}$ reference angle families
- express the value of sine, cosine, and tangent for x, $\pi \pm x$, and $2\pi - x$ in terms of their values of x, where x is any real number
- use the unit circle to explain symmetry (odd and even) and periodicity of trigonometric functions
- choose trigonometric functions to model periodic phenomena for specified amplitude, frequency, and midline
- prove and apply trigonometric identities

Reproducibles are available in a downloadable, printable format. See page xx for directions about how to access them.

Materials

- Unit Circles (Reproducible 30)
- Abe's and Bea's Calculator Keystrokes (Reproducible 31)
- Special Right Triangle Relationships (Reproducible 32)
- Sinusoidal Graph (Reproducible 33)

Good Questions and Teacher Notes (pages 143–153)

1. Suppose that you were walking along a circle with a radius of 1 mile. How could you determine when you had walked an arc of exactly 30 degrees if all you had was a tape measure?

This question asks students to think about angle measure ON the unit circle as opposed from the vertex of the angle at the center of the circle.

2. What trigonometric expressions will have the same graph as $y = \cos x$ for values of x for which it is defined? Generate as many as you can.

This asks students to create their own trigonometric identities. For example, possible solutions include $y = \cos(-x)$; $y = \sin(\frac{\pi}{2} - x)$. Prompt students to find and confirm additional trigonometric identities for expressions like $y = \cos 2x$ and $y = \sin 2x$.

3. Where on the unit circle would a 2 radian angle (arc) be located? Suppose that $2 + x$ represents an arc that shares the same point on the unit circle. What is a possible value of x?

Students often struggle when there are multiple solutions to a question, and this question has infinitely many since students must recognize the periodic nature of radian measure and give an answer that is a multiple of 2π. If no student offers a negative solution, you can discuss how the sign indicates the direction (similar to a number line).

4. Luke, Devon, and Elias each draw an angle (in standard position) on a unit circle. Luke and Devon notice that their terminal rays intercept the unit circle at the same *x*-coordinate. Devon and Elias notice that their terminal rays intercept the unit circle at the same *y*-coordinate. What might their angles look like? What else should the three notice?

Students should notice that each angle has the same reference angle (acute angle formed with the *x*-axis), and that the absolute value of the *x*-coordinates (and *y*-coordinates) will be the same.

You could also ask, *What if their* x- *and* y-*coordinates were reversed?*

5. Students were asked to find *a possible solution* to the equation in the form $a \cos x + b = 0$, where *a* and *b* are real numbers.

What equation might they have been solving and how could all three be correct?

- Abe and Bea got different answers, but when they graphed them on the unit circle, they noticed that the terminal ray of their angles passed through the unit circle in the same place.
- Cia noticed that she got a different answer from both Abe and Bea, and when she graphed it her terminal ray passed through the unit circle in a different place.

A possible answer could be $2\cos x - 1 = 0$, where Abe said $x = -\frac{\pi}{3}$, Bea said $x = \frac{5\pi}{3}$ or $\frac{11x}{3}$, and Cia said $x = -\frac{5\pi}{3}$ or $\frac{7\pi}{3}$.

6. The London Eye (32 capsules on a wheel with a 120-meter diameter) is a 135-meter-tall Ferris wheel that takes thirty minutes to complete a revolution. The Ain Dubai (48 capsules on a wheel with a 210-meter diameter) is a new 250-meter-tall Ferris wheel in Dubai that takes forty-eight minutes to complete a revolution. Both wheels move constantly, slowly enough that passengers can load and unload safely. What are some questions that you can answer using the two models?

This question provides enough information to determine the frequency, midline, and amplitudes for both. Possible models would be:

$$h_{\text{LE}}(t) = 60 \times \sin\left(\frac{\pi}{15}(t - 7.5)\right) + 75$$

$$h_{\text{AD}}(t) = 105 \times \sin\left(\frac{\pi}{15}(t - 12)\right) + 145$$

Students can use the models to find intervals when a rider on one would be higher than a rider on the other. Or they can solve for the time interval where riders on the Ain Dubai would be higher than the max height on the London Eye. They can also determine which Ferris wheel is moving "faster" as passengers load and unload.

7. Students were given the following two diagrams and asked to find (or approximate) the value of sin(45) and cos(45) (Larson and Hostetler 2007, 148).

(See Reproducible 30, Unit Circles.)

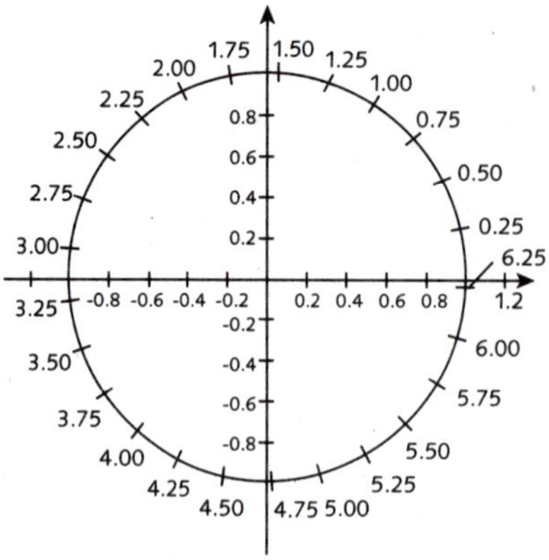

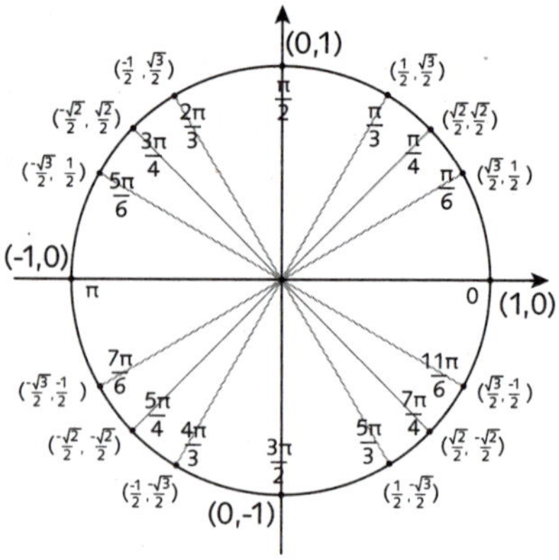

Abe says, using the diagram on the right, that

$\sin(45) = \cos(45) = \frac{\sqrt{2}}{2}$.

Bea says, using the diagram on the left, that

$\sin(45) \approx \sin(1) \approx 0.85$

$\cos(45) \approx \cos(1) \approx 0.5$

Which reasoning is correct? Why?

The 45 in this question could be interpreted as being degrees, in which case the students need to see that $\frac{\pi}{4}$ is the radian equivalent to 45° using the conversion that π radians = 180° when using the diagram on the right. Otherwise, the 45 is in radians and students would need to realize that $45 - 7(2\pi) \approx 1.02$ and use the diagram on the left to approximate the cosine and sine. This emphasizes the importance of using the degree symbol when degrees are your intention. Angles without it are assumed to be in radians.

Ask students to create additional questions that could be answered using either of the two diagrams. For example, they might create an estimate of cos(-5). Or, they may use the diagram on the left to write cos(1), cos(2), cos(3), cos(4), cos(5), cos(6) in order from smallest to largest.

8. For each scenario, is it possible to find a value of x that makes the following expressions true?

- $\sin(x + \pi) = \sin(2\pi - x)$
- $\sin x < \sin(2\pi - x)$
- $\cos(x) < \sin(2\pi - x)$
- $\sin x < \sin(2\pi - x)$ and $\cos(x) < \sin(2\pi - x)$

Create your own scenarios and exchange with a classmate to solve.

This question can be solved using either the unit circle or the graphs of $y = \sin x$ and $y = \cos x$. Encourage students to share their reasoning for solutions and how they created their own scenarios. You can ask students, "Can you generalize a set of values of x for which the statement is true?"

Challenge students to create a scenario for which no solution exists.

9. Abe, Bea, and Cia were checking that $x = 4.1244...$ was a solution to the equation $2 \tan x - 3 = 0$.

What errors were Abe and Bea making?

Abe chose to solve the problem algebraically, getting

$2 \tan x - 3 = 0$

$\tan x = \frac{3}{2}$

$x = \tan^{-1}\frac{3}{2} = 0.982793723247 \neq 4.1244$

And he concluded that 4.1244... is not a solution.

Bea plugged in 4.1244... for x and got $2 \times \tan(4.1244) - 3 = -2.85578209428 \neq 0$ and also concluded that 4.1244 is not a solution.

Cia also plugged in 4.1244 for x and evaluated the expression (after checking what mode her calculator was in) and got $2 \times \tan(4.1244) - 3 = 0.0000885523682532 \approx 0$ and concluded that 4.1244... was indeed a solution.

This question reminds students that they must be sure that they use the correct mode (degree versus radian) when using technology and that the range of the inverse tangent is $-\frac{\pi}{2}$ to $\frac{\pi}{2}$.

10. For the following, let *a* represent a real number.

Abe and Bea were typing in the following sets of keystrokes on their calculators.

Abe thinks that the result will always yield the value of *a*. Bea disagrees and says that only three combinations of keystrokes *always* yield the value of *a* for values of *a* for which the expression is defined.

(See Reproducible 31, Abe's and Bea's Calculator Keystrokes.)

Set I	Set II
$\boxed{\tan^{-1}}\ \boxed{\tan}\ a$	$\boxed{\tan}\ \boxed{\tan^{-1}}\ a$
$\boxed{\sin^{-1}}\ \boxed{\sin}\ a$	$\boxed{\sin}\ \boxed{\sin^{-1}}\ a$
$\boxed{\cos^{-1}}\ \boxed{\cos}\ a$	$\boxed{\cos}\ \boxed{\cos^{-1}}\ a$

Which ones don't always yield the value of *a*, and for what values of *a* will this happen? Check by evaluating the expressions in certain values for *a* and observing how the domains need to be restricted so that all 6 are true.

Here, students must see that the domain of sine, cosine, and tangent must be restricted in order for an inverse function to exist. In Set I, $0 \le a \le \pi$ for cosine and $-\frac{\pi}{2} \le a \le \frac{\pi}{2}$ for sine and tangent. For Set II, $-\infty < a < \infty$ for tangent and $-1 \le a \le 1$ for sine and cosine.

Ask students to discuss why this must be true and also what other possible restrictions *could* have been used.

11. After learning that:

- an *even* function can be defined as any function such that $f(-x) = f(x)$ or as a function that has symmetry about the y-axis
- an *odd* function can be defined as any function such that $f(-x) = -f(x)$ or as a function that has symmetry about the origin

Abe wondered if the sine, cosine, and tangent functions were even, odd, or neither. He drew a unit circle and the graph of each function and quickly determined the answer. What did he draw and what were his conclusions?

Here, Abe likely drew something like:

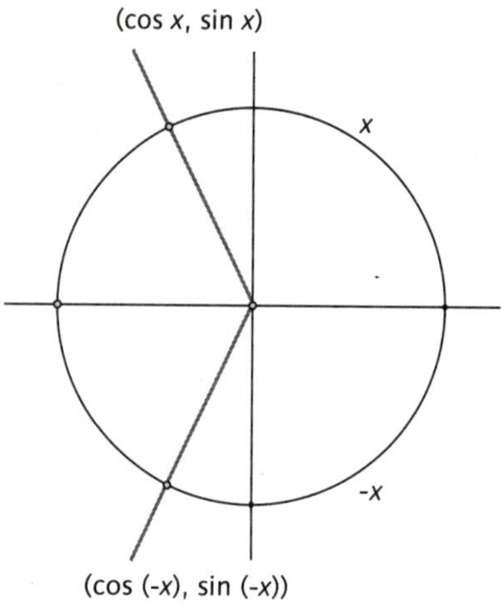

and noted that the two cosine values were the same, the sine values were opposites, and that the terminal rays had opposite slopes; thus the cosine is even and the sine and tangent are odd functions. Similarly, the symmetry around the y-axis can be seen looking at the graph of the cosine function, and the symmetry around the origin can be seen in the graphs of the sine and tangent functions.

12. GQ drew a unit circle with a terminal ray drawn in the first quadrant subtending an angle of x radians. She labeled the point where the terminal ray intersected the unit circle (u, v) using the definition of sine and cosine; that is $(\cos x, \sin x)$. Then she drew a few segments in the figure and noticed something interesting:

$$\sin^2 x + \cos^2 x = 1$$

What did her diagram look like? And did she need to draw the terminal ray in the first quadrant?

This leads students to notice the Pythagorean identity.

13. Abe and Bea were solving two equations:

$$\sin^2 x - \cos^2 x = 0 \qquad \text{and} \qquad 2\sin^2 x - 1 = 0$$

After getting the same solution to both, $x = \frac{\pi}{4}, \frac{3\pi}{4}, \frac{5\pi}{4}, \frac{7\pi}{4}, \ldots$, they wondered if $\sin^2 x - \cos^2 x = 2\sin^2 x - 1 = a$ for any $-1 \le a \le 1$.

How could you demonstrate (prove) that this is true?

Here students can use the Pythagorean identity to prove this identity, or they could show that the graph of both expressions is the same, which is also why a is restricted to [-1, 1].

Ask students to solve the original equations for $a = \frac{1}{2}$. This shows that one version of the identity is more useful in certain circumstances, in this case the $2\sin^2 x - 1$ version. Also, you could introduce another equivalent expression $-\cos 2x$ and discuss how that could be used when $a = \frac{1}{2}$.

14. Recall from special right triangles the following relationships.

(See Reproducible 32, Special Right Triangle Relationships.)

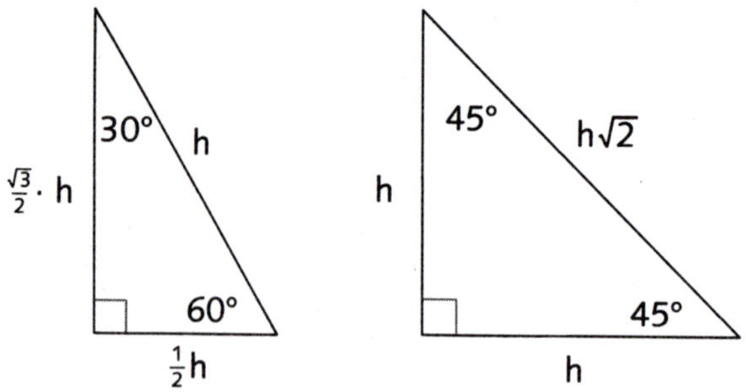

GQ claims that she can utilize these two diagrams and a unit circle to evaluate any of the following.

Explain GQ's thinking.

- $\sin\frac{5\pi}{6}$
- $\cos\frac{5\pi}{4}$
- $\tan\frac{5\pi}{3}$

This allows students to see the "associated" right triangle for any trigonometric expression. Thus circular functions are an extension of right triangle trigonometry rather than an entirely new construct.

15. Determine as many possible equations that describe the following graph.

(See Reproducible 33, Sinusoidal Graph.)

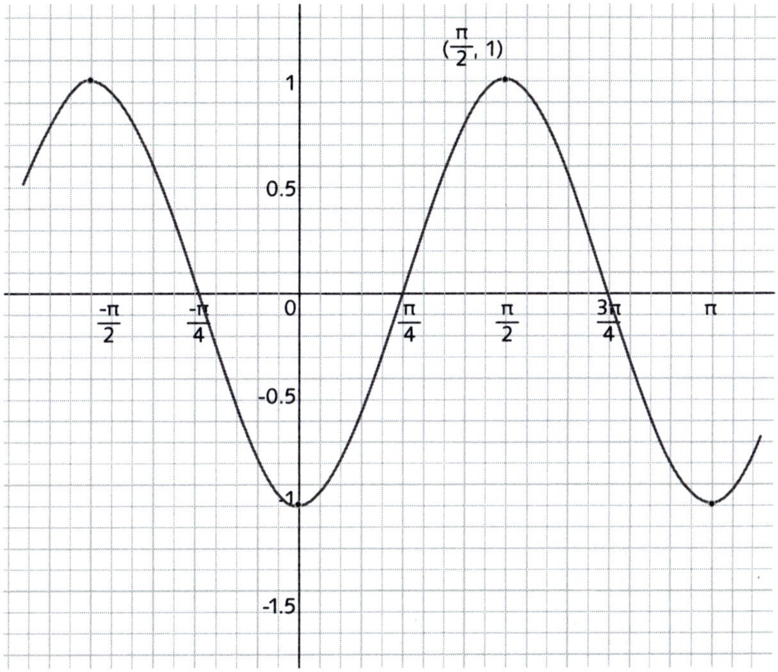

The graph has several possible equations. Students will likely find $y = -\cos(2x)$. The more obscure solutions are $y = 2\sin^2 x - 1$ and $y = \sin^2 x - \cos^2 x$, which can lead to a discussion of trigonometric identities.

Inverse Functions and Composites

Experiences at This Level Will Help Students To

- understand that the inverse of a function is not necessarily a function—and thus the domain of the original function is limited so that its inverse will be a function
- decompose functions into composite functions
- connect evaluating composite functions to evaluating a chain of functions

Reproducibles are available in a downloadable, printable format. See page xx for directions about how to access them.

Materials

- Composing Functions (Reproducible 34)

Good Questions and Teacher Notes (pages 155–159)

1. **Is it possible to have two functions that compose to allow the initial input to always equal the final output? That is, can you find f and g so that $f(g(x)) = x$? If it is possible, create a pair to demonstrate. If it is not possible, explain why not.**

This is possible whenever f and g are inverses. This question leads to great group discussions after composite functions are introduced but before students learn about inverses. Students should generate a few different types of inverse functions (likely linear) that will enable you to have them think about why those pairs are inverses.

2. **Ameris has a problem—she doesn't know which expression below is $f(x)$ and which is $g(x)$. What she *does* know is that $f(g(3)) = 7$. Figure out which functions can be f and g. Is there more than one answer?**

- $3x - 5$
- $(x - 1)^2$
- x
- $\sqrt{x + 1} + 1$

This question has many answers. However, from the given information, students should recognize that x could be composed with $3x - 5$ and $(x - 1)^2$ to get the desired result. In addition, since $\sqrt{3 + 1} + 1 = 3$, then $\sqrt{x + 1} + 1$ can also be composed with $3x - 5$ and $(x - 1)^2$ to get the desired result.

3. Assume that $f(x) = 2x + a$ for some a value. If $g(x) = x^2 + 4$, then how does the graph of $h(x) = f(g(x))$ change as a changes? Explain. How about the graph of $k(x) = g(f(x))$?

Students should note that in the case of $h(x) = f(g(x))$, $g(x) = x^2 + 4$ only impacts the input of $f(x)$. Thus, as a changes, $f(g(x))$ moves vertically about the cartesian plane. In the case of $k(x) = g(f(x))$, however, a impacts the input of $g(x)$. This means that as a changes, the graph of $g(f(x))$ moves both vertically and horizontally in the cartesian plane.

4. Which functions below are inverses? How do you know?

a. $y = \frac{2}{3}x + 4$

b. $y = \frac{3}{2}x - 4$

c. $y = \frac{3}{2}x - 6$

d. $y = 4x + \frac{2}{3}$

e. $y = \frac{2}{3}(x + 4)$

f. $4y = 2x - 3$

Students can tackle this question a variety of ways: algebraically, graphically, testing points, etc. This question gets at the visual features of inverses. We recommend that you ask students: *How would you know those were inverses before composing them?* This is an opportunity for students to build structural insight into the relationship of inverses.

5. **Do you need functions in equation form to compose? For example, examine functions $f(x)$ and $g(x)$ in the following. What is $f(g(x))$? Show both in table and graph form.**

(See Reproducible 34, Composing Functions.)

x	$f(x)$
-3	26
-2	17
-1	10
0	5
1	2
2	1
3	2
4	5
5	10

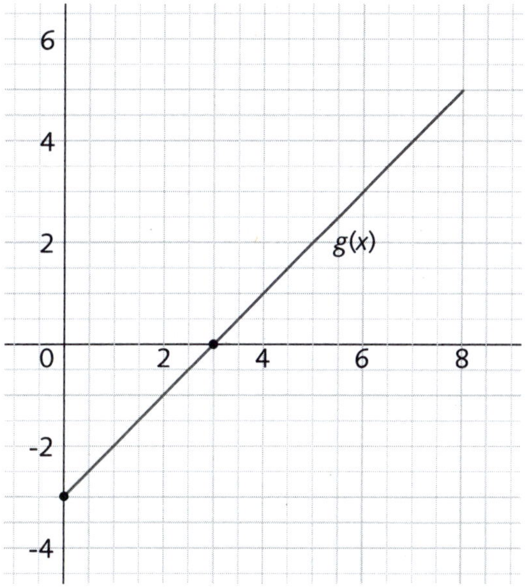

The point of this question is to reorient students' perceptions of the input/output of functions as they compose. Since the graph of $g(x)$ is given for $0 \le x \le 8$, these are the x-values they should start with. Have students work together on recognizing how the output of $g(x)$ is the input of $f(x)$. Questions you might ask are, *What was the effect of* g(x)*? What did it do to* f(x)*?* Students might see that $g(x)$ translated the points of $f(x)$ to the right 3 units.

6. What two functions f and g can be composed to become $h(x) = 3 + \sqrt{2(x-3)^2 + 1}$? Can you find more than one pair? Be ready to share.

This is an open-ended question that allows students to select a variety of functions that could compose to become $h(x)$. For example, if $f(x) = (x - 3)^2$, then $g(x) = 3 + \sqrt{2x + 1}$. Or if $f(x) = \sqrt{2(x-3)^2 + 1}$, then $g(x) = 3 + x$. If your students are stumped, you may want to start with an easier version first, such as $k(x) = \frac{x-2}{5} + 4$.

7. Is the inverse of $f(x) = (x - 1)^2$ a function or not? Explain. If it is not a function, how can the domain of $f(x)$ be restricted so that its inverse is a function?

This question gets students to start recognizing that the inverse of functions is not always a function. Students should recognize that if the domain of $f(x)$ is restricted to either $x \geq 1$ or $x \leq 1$, then its inverse is a function. You might want to use this opportunity to have students recognize that the new domain of $f(x)$ is related to the range of its inverse.

8. Kelly says that if (x_1, y_1) is a point on a function $f(x)$, then (y_1, x_1) must be a point on its inverse. Do you agree? Why or why not?

Kelly is correct as long as the point (x_1, y_1) is included in any restricted portion of $f(x)$ (in the case where the inverse of $f(x)$ is not a function). If your students accept this statement as true without recognizing its limitations, ask about $f(x) = x^2$ and the point (-1, 1). Ask, "Can you alter Kelly's statement so that it is always true?"

9. What function is its own inverse? Is there more than one answer?

This is tricky, but typically after thinking about it a bit, students will figure out that $f(x) = x$ is its own inverse. In addition, however, any line $y = -x + a$ will be its own inverse.

CHAPTER 7

Good Questions for Geometry

Students enter high school with varying levels of familiarity with geometric objects and concepts. In high school they more intentionally and formally analyze characteristics and properties of two- and three-dimensional geometric shapes and develop mathematical arguments about geometric relationships; they connect that analysis to specify locations and describe spatial relationships using coordinate geometry and other representational systems; they apply transformations and use congruence, similarity, and symmetry to analyze mathematical situations; and they use visualization, spatial reasoning, and geometric modeling to solve problems including trigonometry (NCTM 2016). Finally, these concepts are applied to support the study of functions and pre-calculus.

Measurement

Experiences at This Level Will Help Students To

- appreciate that units matter
- understand that spatial dimensions, such as base and height, must be perpendicular
- make sense of a measurement as representing a quantity of an iterated unit of measure without gaps or overlaps
- understand that there is an inverse relationship between the number of units needed to measure an object and the size of the unit
- understand that all physical measurement is approximate
- understand that the area and perimeter of an object are not necessarily directly linked (e.g., increasing perimeter does not necessarily mean that area is changing and vice versa; however, sometimes they are [such as in a circle or square])
- measure by decomposing a shape into parts and finding the sum of all parts or by decomposition (i.e., finding the length, area, or volume of a larger known and measurable object and then removing a part)

Reproducibles are available in a downloadable, printable format. See page xx for directions about how to access them.

Materials

- Marisol's Pen (Reproducible 35)
- Geometric Shapes A, B, C, and D (Reproducible 36)
- Four Triangles (Reproducible 37)
- Sector of a Circle (Reproducible 38)

Good Questions and Teacher Notes (pages 163–168)

1. **Cali said her triangle has an area of 10 while Eddie says his triangle has an area of 8. Is Cali's triangle necessarily bigger? Explain.**

This question raises a few potential issues for students to discuss. For example, since neither student provided the units with which their areas were measured, then Eddie's triangle could have a larger area if his square unit is sufficiently larger (for example, if Eddie's was measured in square inches and Cali's in square centimeters). In addition, when the question is about which triangle is "bigger," there is ambiguity about the quantity upon which that decision should be made. Some students might argue that since the areas are discussed, the comparison should be based on area. However, other students might feel that the height or perimeter of the triangles is a quantity that most matters.

2. Marisol wants to measure the length of the pen next to the ruler in this photo. What is a good estimate of its length? Explain how you got your answer.

(See Reproducible 35, Marisol's Pen.)

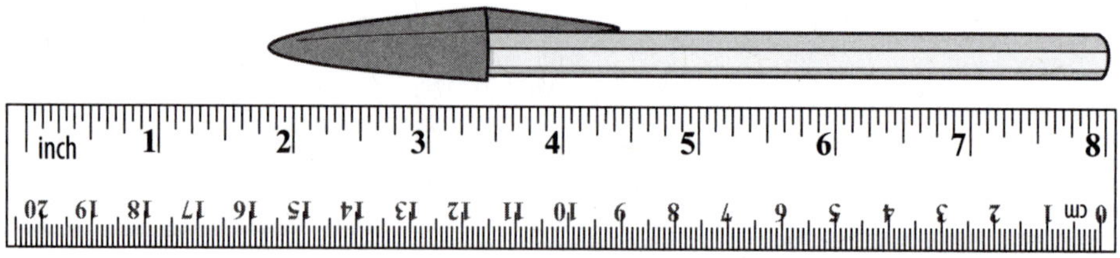

This question brings up the need to understand length as a number of iterated units. Students might say "seven inches" if they are only counting the number of inch *markers* (that is, counting the numbers 2, 3, 4, 5, 6, 7, and 8) as opposed to "six inches" if they are counting full iterated inches. Ask students to show you where one inch is and help them recognize it is the length between the 2 and 3 markings as opposed to at a single marking. This question also will require students to deal with partial inches.

3. **Compare the following geometric shapes. Assume each has equal width and height overall. Rank the objects from smallest to largest based on their areas. Would their perimeters get the same ranking? Explain how you know.**

(See Reproducible 36, Geometric Shapes A, B, C, and D.)

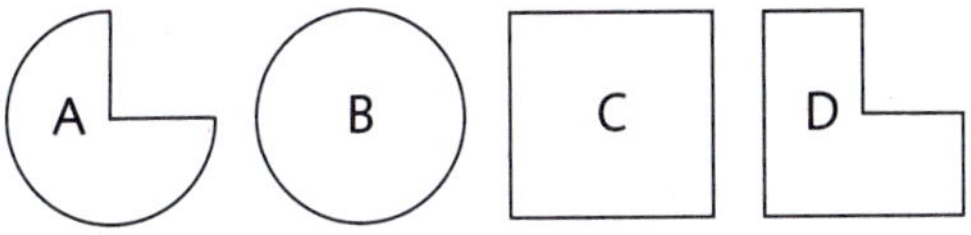

This question requires students to grapple with the difference between area and perimeter. It also requires them to consider what they know about squares and circles. Most students will immediately recognize that A has less area than all of the others and that C has the most area (in fact, C can contain each of the others with room to spare). However, they may be puzzled about how to decide which has larger area between B and D. Here, recognizing that B has area of πr^2 square units while D has area of $3r^2$ square units will enable students to recognize that B has a larger area than D. The perimeters will be ranked differently. Students might be surprised to see that C and D have equal perimeter (of $8r$ units), and both are larger than B (with $2\pi r$ units, or approximately $6.28r$ units) and A (which is $\frac{3}{4}2\pi r + 2r \approx 6.71r$ units).

4. For each expression that follows, draw a trapezoid having an area that could be calculated this way. Explain how your expression represents the area of the trapezoid.

$\frac{1}{2}(3)(4) + (9)(4) + \frac{1}{2}(5)(4)$

$(17)(4) - \frac{1}{2}(3)(4) - \frac{1}{2}(5)(4)$

$\frac{1}{2}(9)(4) + \frac{1}{2}(17)(4)$

This question requires students to consider how a geometric object, such as a trapezoid, can be decomposed into parts, and how its area is the sum of those parts. Each of these expressions can represent the same trapezoid with bases of 9 and 17 units and a height of 4 units.

5. Compare the areas of the four triangles in the following diagram. Explain your reasoning.

(See Reproducible 37, Four Triangles.)

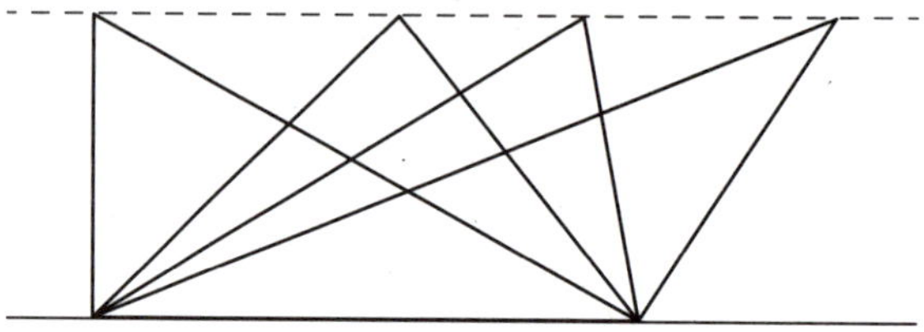

This question provides an opportunity for students to realize that all triangles with equal base and height have equal area. Therefore, when finding the area of a triangle, they are able to share one vertex along a line parallel to the base of the triangle and the area will not change.

6. Find the area of this shape using two different strategies.

(See Reproducible 38, Sector of a Circle.)

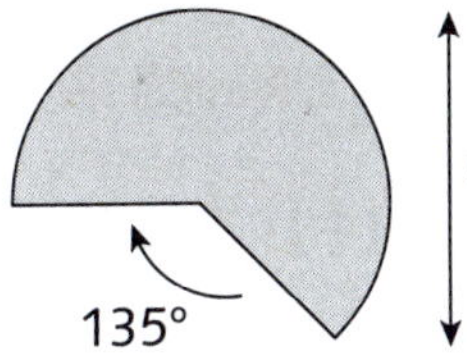

Students may approach this problem in several ways, depending on how they decompose the region. They may think of it as a whole circle with a part removed, or as the composition of two or more regions. Also, student may mistakenly assume that the radius of the circle is 4 units.

7. When carefully walking heel-to-toe directly across the classroom, Marianna says that the width of the classroom is 62 "feet" while Marissa says that it is 56 "feet." Can they both be right? How? If so, what observations can you make? What can the two students do about the difference?

Students should recognize that for both to be correct means that Marianna and Marissa have different-size feet, and that Marissa's feet must be longer since it took fewer of them to span the width. Students should recognize that selecting a unit that spans the room will resolve the issue. Often floor or ceiling tiles will work. Or, students should recognize that this issue is a reason why standard units, such as meters or inches, were developed.

8. **Ms. Geo has bins in her classroom in which she stores her 1-inch cubes. Each bin is shaped like a prism. When packed into the bin, she can get 18 layers of cubes, where each layer is a solid 12-by-22-inch rectangle of cubes. If she has 6 full bins of 1-inch cubes, how many 1-inch cubes does she have?**

This question supports the understanding of volume of a prism as the sum of multiple identical layers, where each layer has a height of 1 unit and an area equal to the base area of the prism.

Congruence and Similarity

Experiences at This Level Will Help Students To

- recognize congruence as a special case of similarity; that is, if two figures are similar with a ratio of 1, then they are congruent
- use the relationship between geometric objects to deduce properties of each geometric object
- understand that when geometric objects are dilated, the image is similar to the original object

Reproducibles are available in a downloadable, printable format. See page xx for directions about how to access them.

Materials

- Figures A, B, C, D, and E (Reproducible 39)
- Overlapping Triangles (Reproducible 40)
- ΔABC and ΔDCB (Reproducible 41)
- Three Right Triangles (Reproducible 42)
- Two Versions of 30-60-90 Triangles (Reproducible 43)

Good Questions and Teacher Notes (pages 170–176)

1. If $\Delta ABC \cong \Delta DEF$ and $\Delta ABC \cong \Delta EFD$ and $\Delta ABC \cong \Delta FDE$, what can you say about ABC? Explain. Use diagrams to show your thinking.

This question draws students' attention to the naming conventions of congruence. If the two triangles are congruent no matter what order you name the vertices, then both triangles must have rotational symmetry about their center, and therefore must be equilateral.

2. Raul dilated quadrilateral $ABCD$ by a factor of 2 from the point (1, 2). Randall dilated the same quadrilateral $ABCD$ by a factor of 2 from the point (5, -1). How are the resulting dilations the same or different? Explain your thinking.

This question encourages students to recognize that the "stretch point" or point from which a geometric object is dilated changes the location of the resulting dilation. However, the size of the dilation depends on the stretch factor. Therefore, Raul's and Randall's resulting dilations must be congruent. In addition, the corresponding sides of the dilations will be parallel.

3. Assume point *A* lies on some line *l*. If $\triangle ABC$ is symmetric with respect to line *l*, what statements can you make about $\triangle ABC$?

This question uses symmetry to help students recognize congruent properties of a triangle. Specifically, students may note that $\triangle ABC$ must be isosceles (or, more specifically, it could be equilateral). Also, students may point out that *l* must intersect side *BC* at its midpoint. If this midpoint is labeled *M*, then $\overline{BM} \cong \overline{MC}$. Angles *B* and *C* must be congruent to each other as well. You may want to have students collaborate on this question with tracing paper, which allows them to construct the triangle.

4. Examine the following figures. Which are similar? Which are congruent? How can you tell?

(See Reproducible 39, Figures A, B, C, D, and E.)

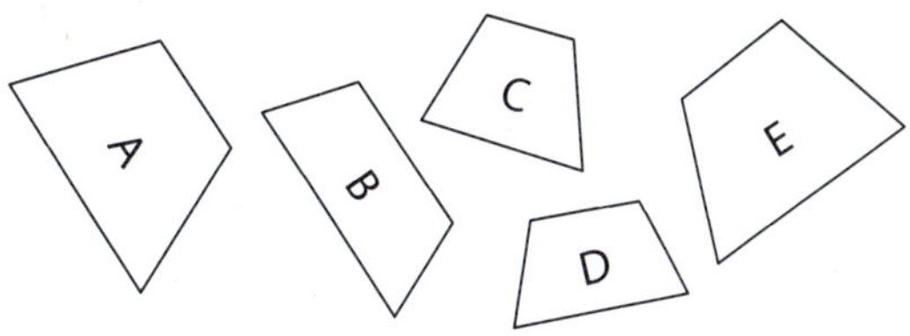

This question challenges students to recognize proportionality and to understand that similarity is not dependent on the orientation of the figure. Solicit ideas for how students could determine if shapes, such as A and C, are similar. Students may want to use trace paper to test for congruence or to see if angles have the same measure. Ask students what should be true about the side lengths if the figures are similar. Students may argue that figures A and E are congruent, but they can also be described as similar with a ratio of 1.

5. Penny and Percy both constructed a triangle with one side with length 20 units and another side with length 10 units. When they measured the angle opposite the side with length 10 units, they got the same measure too! However, their triangles are not congruent. How can that be? Show your thinking on a diagram.

This question brings up the ambiguous case (SSA). Students should draw a diagram to show how two triangles can result for angles < 30°. A possible diagram is shown here.

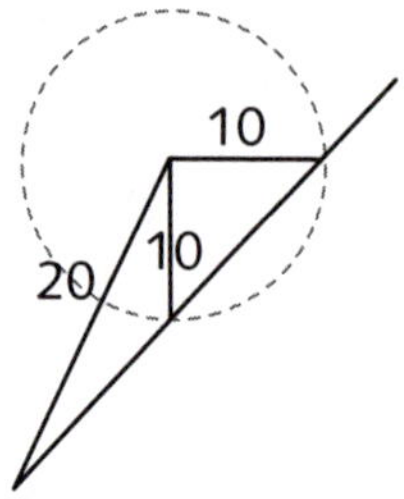

6. Examine the following diagram, showing a triangle with two sides lengths (32 units and 17 units) and a 20° angle that is not between the two sides. Note that there could be two possible triangles for these measures. For what acute angle (instead of 20°) would there only be one possible triangle? No triangles? Show how you know.

(See Reproducible 40, Overlapping Triangles.)

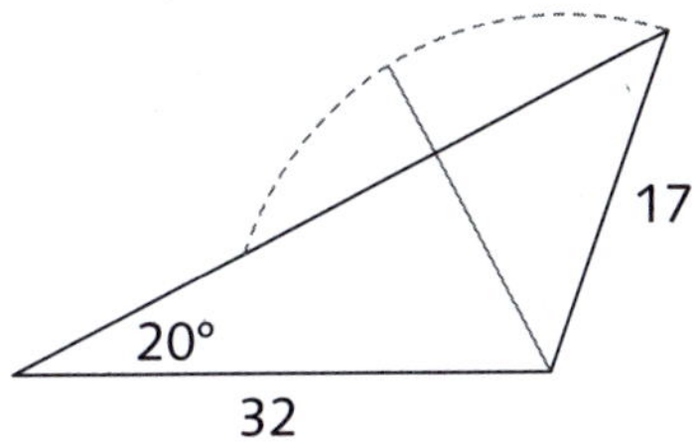

This question challenges students to recognize that there will only be one possible triangle in the case that $\tan(\alpha) = \frac{17}{32}$, and $\alpha \approx 28°$. Then for all angles greater than 28°, no triangle is possible.

7. If $\triangle ABC \cong \triangle DCB$ and $DBC = 40°$ and $DCA = 55°$, then what true statements can you make about the sides, angles, and figures in this diagram? (See Reproducible 41, ΔABC and ΔDCB.)

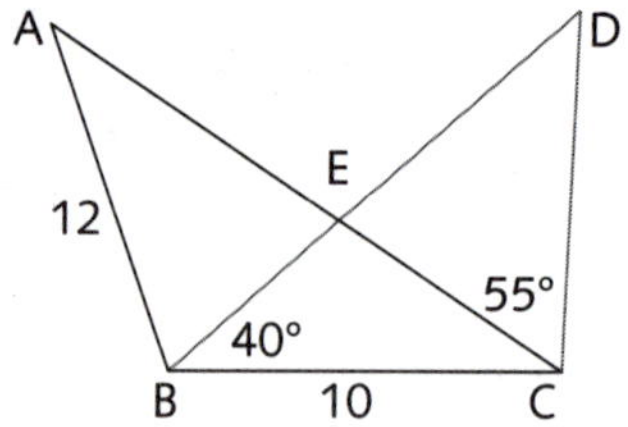

This question challenges students to recognize congruent triangles that overlap and also prompts them to recognize which corresponding parts must be congruent. A variety of other observations can be made, such as how $\triangle BEC$ is isosceles. Students may also use trigonometric tools if available to find lengths AC and BD, as well as the lengths of BE and CE.

8. Millie and Steven each constructed a triangle. Each triangle has one angle with measure 24°, a side with length 10 cm, and another side with length 8 cm. Assuming that their measures are accurate, are their triangles congruent? Why or why not?

This question can generate a range of discussion topics based on students' level of understanding of congruent triangles. Early in their learning, this problem can provoke students into constructing triangles themselves with these measures to determine if more than one triangle is possible. (It is.) Later, once students understand congruent theorems, this problem can provoke a discussion on the various ways two sides and an angle can be arranged. A question you might ask during any discussion is *What other information would help you know if the triangles are congruent and why?*

9. Solve for the lengths marked x, y, and z in this diagram. What relationships support your conclusions?

(See Reproducible 42, Three Right Triangles.)

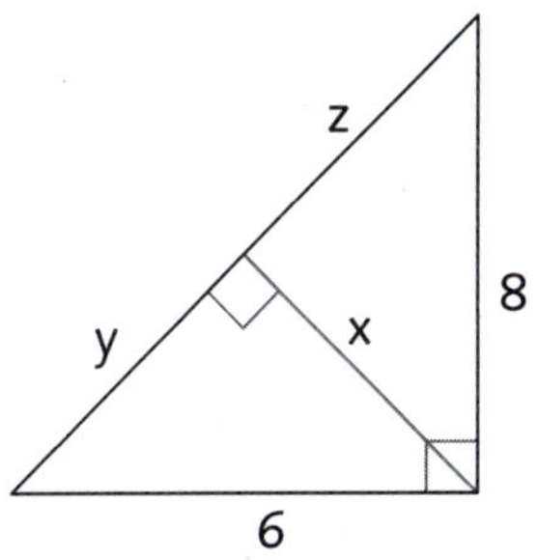

Students can use a variety of strategies for this problem—and if they all use the same strategy, this may indicate that you should focus on emphasizing multiple ways of solving problems in your class. For example, students can use the Pythagorean theorem to solve for the length of $y + z$ (which is 10 units). Then, they can use the fact that the three triangles are similar to create ratios that can help solve for some of the variables. An example for this is $\frac{x}{6} = \frac{8}{10}$, leading to $x = 4.8$ units. Another approach is to recognize that the area of the triangle remains the same whether using 6 and 8 units for base and height or using 10 and x units for base and height. That means that both areas must be equal, leading to $\frac{1}{2}(6)(8) = \frac{1}{2}(10)(x)$. Further use of the Pythagorean theorem leads to $y = 3.6$ and $z = 6.4$ units.

10. **Explain how to justify that the following two triangles are congruent using at least two different strategies. Can you think of more?**

(See Reproducible 43, Two Versions of 30-60-90 Triangles.)

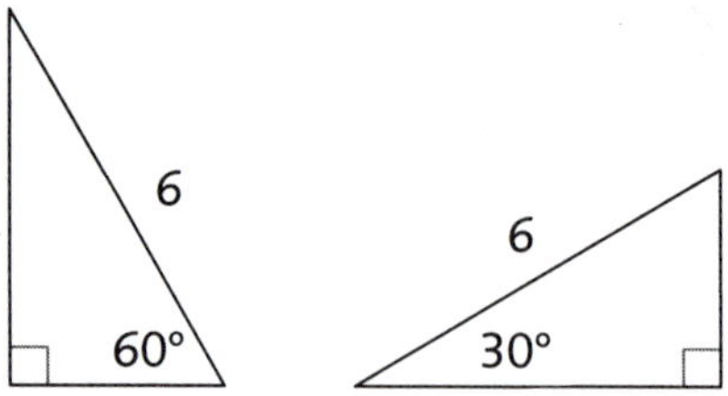

This problem prompts students to use multiple relations (and not just the most obvious ones) to determine congruence for a pair of triangles. Expect some students to find AAS relatively quickly and use the triangle sum to determine ASA. But special right triangle relationships or trigonometry can help them determine the length of one or both legs of each triangle, allowing for students to use SSS, SAS, and HL as well.

TRANSFORMATIONS

EXPERIENCES AT THIS LEVEL WILL HELP STUDENTS TO

- apply transformations and use symmetry, congruence, and similarity to analyze mathematical situations
- use various representations to help understand the effects of simple transformations and their compositions
- understand and represent translations, reflections, rotations, and dilations of objects in the plane by using sketches, coordinates, and function notation

Reproducibles are available in a downloadable, printable format. See page xx for directions about how to access them.

MATERIALS

- ΔABC and ΔDFE (Reproducible 44)
- Rotating Quadrilateral (Reproducible 45)
- ΔABC and Point D (Reproducible 46)
- ΔABC (Reproducible 47)

GOOD QUESTIONS AND TEACHER NOTES (PAGES 178–184)

1. Can a sequence of translations always be replaced with a single translation? Why or why not?

Yes, it can. Ask students to provide examples to support their answers. Also you can extend this question by asking students whether a single translation can be replaced by a sequence of reflections.

2. What is a sequence of transformations that moves ΔABC to ΔDFE? Generate at least two distinct ways.

(See Reproducible 44, ΔABC and ΔDFE.)

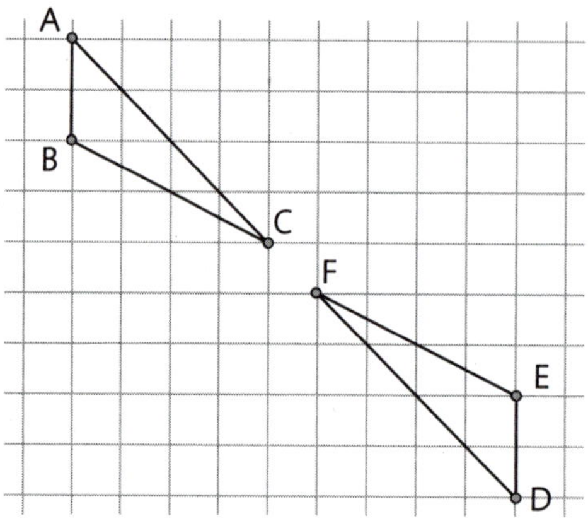

This question can help students realize that sometimes a rotation can be achieved by a combination of a rotation and translation (such as rotating about *C* 180° and then translating *C* to *F*) or by two carefully chosen reflections (such as reflecting across a vertical line halfway between *C* and *F* and then reflecting the image across a horizontal line halfway between *C* and *F*).

3. What solid would be created by rotating the shaded region about the vertical pole? Sketch the solid on your paper. Then design a new region that would create a different solid.

(See Reproducible 45, Rotating Quadrilateral.)

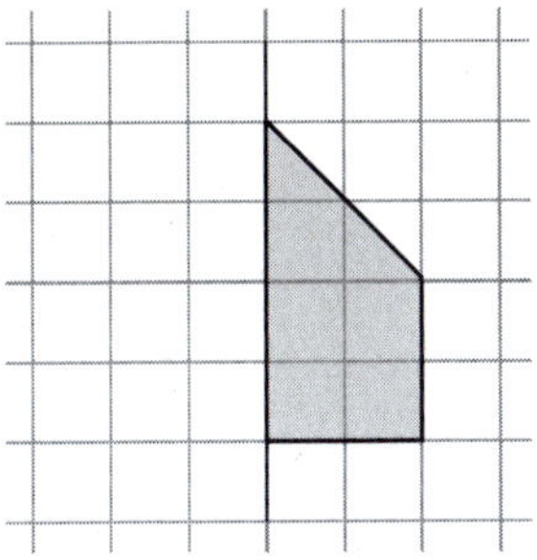

This question helps students to visualize solids of revolution. You can also provide a solid and ask students to state what shaded region and axis of rotation could be used to create it.

4. Can the order of a sequence of transformations be changed without changing the image? If so, under what conditions? When can it not be changed?

Yes, this is possible. For example, the order of a sequence of translations is commutative. However, a sequence of reflections may not be.

5. When can a regular polygon be tiled? Explain.

Students need to notice that the interior angles must be a factor of 360.

6. When can two regular polygons be tiled? Explain.

Students need to notice that the sum of interior angles must be 360°. For example, students may notice that two dodecagons and a triangle (150 + 150 + 60 = 360) will tile the plane.

7. If all points (x, y) of a graph of $f(x)$ are changed through the sequence of transformations as indicated here, what is the resulting transformation? Explain your thinking.

$(x, y) \rightarrow (x, -y)$

Then for the image $(x, y) \rightarrow (y, x)$

Then for the image $(x, y) \rightarrow (y, -x)$

This question may result in a surprise for students as the original function must return to its original state. An extension to this question could be to ask students to generate their own "returning" set of transformations.

8. If all points (x, y) of a graph of $f(x)$ are changed to $(-x, -y)$, what is the resulting transformation? Explain your thinking.

Students should notice that the result is a 180° rotation about the origin.

9. What changes with rotation? What stays the same?

Here students should notice that rotations and translations preserve congruence and orientation, while reflections preserve congruence but reverse the orientation. This is why a sequence of two reflections over parallel lines results in a translation, and two rotations over non-parallel lines results in a rotation. You can extend this question by changing rotation to reflection or translation.

10. What changes with dilation? What stays the same?

Students should note that dilations preserve similarity and orientation.

11. **Given a triangle in the coordinate plane whose vertices are lattice points (i.e., integer coordinates), generate at least two different reflections over a line that is neither vertical nor horizontal (i.e., $y = mx + b$, $m \neq 0$) such that the image has vertices that are lattice points.**

You may want to scaffold this by providing the coordinates of the vertices of the preimage. Any line parallel to $y = x$ will work. Can students find other solutions?

12. **Given a triangle in the coordinate plane whose vertices are lattice points (i.e., integer coordinates), generate at least two different rotations such that the image has vertices that are lattice points. Does the center of rotation matter?**

You may want to scaffold this by providing the coordinates of the vertices of the preimage.

Ask students if they can generalize a method. For example, students might say that rotating any multiple of 90° around any lattice point will work.

13. Consider the triangle $\triangle ABC$ and point D shown here.

Create a transformation or sequence of transformations so that point D is in the interior of the image.

(See Reproducible 46, $\triangle ABC$ and Point D.)

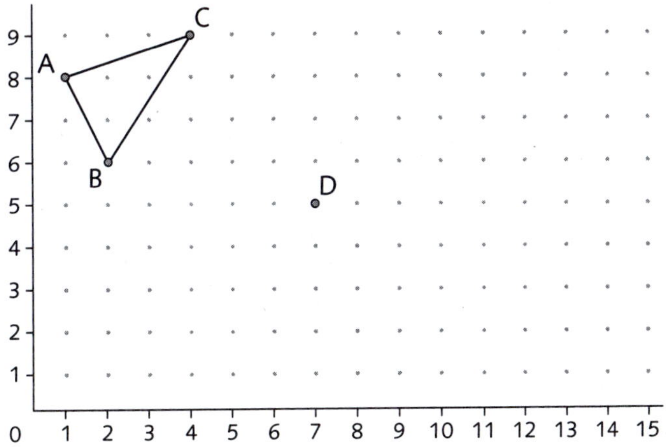

This question allows students to explore various transformations. Encourage students to share their thinking. For example, one sequence of transformations would be a horizontal translation of 5 units (right) followed by a reflection over the horizontal line $y = 6$.

14. Consider the triangle ΔABC shown here.

Create a dilation such that the image has a perimeter that is half that of the preimage. Can you create a dilation such that the area of the image is half the area of the preimage? Does the center of the dilation matter?

(See Reproducible 47, ΔABC.)

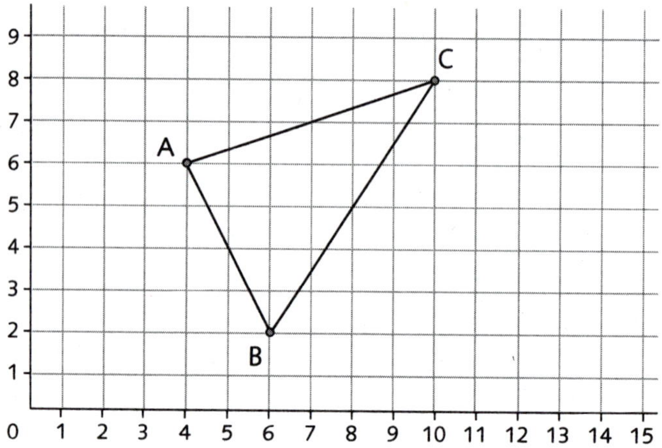

Here students need a scale factor of $\frac{1}{2}$ to accomplish the first goal, and a scale factor of $\frac{\sqrt{2}}{2}$ to accomplish the second. The center of the dilation does not matter, though each produces a different yet congruent image.

Polygons

Experiences at This Level Will Help Students To

- understand that the area of regular polygons can be calculated using the area of a right triangle formed by the apothem, radius, and half a side length
- generalize that the sum of the interior angles of a polygon must be $180(n - 2)°$
- generalize that the central angles of a polygon sum to 360°

Reproducibles are available in a downloadable, printable format. See page xx for directions about how to access them.

Materials

- Square *ABCD* and Δ*EGF* (Reproducible 48)
- Linear Angles (Reproducible 49)
- Square Cut by Two Perpendicular Lines (Reproducible 50)

Good Questions and Teacher Notes (pages 186–192)

1. **What is the relationship between the areas of the regular triangle and hexagon if both have a radius (the length from the center of the polygon to each of its vertices) of 1 unit? Explain.**

The area of the equilateral triangle with radius 1 is $\frac{3\sqrt{3}}{4}$ square units and the area of the regular hexagon with radius 1 is $\frac{3\sqrt{3}}{2}$ square units. Therefore, the regular hexagon has twice the area. This relationship can be recognized without calculating the areas, however. For the equilateral triangle and the regular hexagon with equal radius, the right triangle formed with the radius and apothem are congruent. That is, each triangle is a 30-60-90 triangle with side lengths $\frac{1}{2}$, $\frac{\sqrt{3}}{2}$, and 1 unit. So these triangles must have equal area. Since the equilateral triangle is composed of 6 of these right triangles, while the regular hexagon is made up of 12 of these right triangles, the regular hexagon must have an area that is twice the equilateral triangle.

2. **Is it possible for the measure of one angle of a triangle to be the sum of the measures of the other two angles? If so, give an example. If not, explain why not.**

This is possible. In fact, it's always true for all right triangles; the two acute angles always have a sum of 90°, which is the measure of the third angle.

3. If the measure of a central angle of a regular polygon is an integer, what are its possible measures?

This question challenges students to recognize that a central angle cannot take on certain values (such as 0° or 180°) and that to be an integer, it must be a factor of 360°. So possible measures are 1°, 2°, 3°, 4°, 5°, 6°, 8°, 9°, 10°, 12°, 15°, 18°, 20°, 24°, 30°, 36°, 40°, 45°, 60°, 72°, 90°, and 120°.

4. Estelle is thinking of a quadrilateral that has two diagonals that are perpendicular. What could her quadrilateral be? Justify your conclusion.

This question encourages students to mentally construct quadrilaterals from their diagonal properties. With diagonals that are perpendicular, the resulting quadrilateral could be a square (in the case where they bisect each other and are equal lengths), more generally a rhombus (in the case where they bisect each other), more generally a kite (in the case where only one diagonal is bisected, regardless of whether the diagonals are equal lengths or not), or an isosceles trapezoid (in the case where the diagonals are equal lengths and subdivide each other in equal proportions).

5. For which regular polygons is the central angle equal to the interior angle? Justify your conclusion.

A regular polygon with n sides has a central angle with measure $(\frac{360}{n})°$ and an interior angle with measure of $(\frac{180(n-2)}{n})°$. These are equal only when $180(n - 2) = 360$ or $n = 4$. So, the square is the only regular polygon with this special property.

6. Ari has drawn a regular polygon that has an interior angle with a measure of 108°, while Kathryn's regular polygon has a central angle measure of 72°. Which student's polygon has more sides? How do you know?

They have the same number of sides because regular pentagons have central angles of $360° \div 5 = 72°$ and an interior angle of $180°(5 - 2) \div 5 = 108°$.

7. What is the relationship between the opposite angles of an isosceles trapezoid? How do you know?

The opposite angles of an isosceles trapezoid must be supplementary. Consider an isosceles trapezoid $ABCD$ where $\overline{AB} \parallel \overline{CD}$. The consecutive angles A and D are same-side interior angles to parallel lines, and therefore are supplementary. Also $\angle D$ and $\angle C$ are base angles of the isosceles trapezoid, so therefore must be congruent. So if $\angle A$ and $\angle D$ are supplementary and $\angle D$ and $\angle C$ have equal measure, then $\angle A$ and $\angle C$ must be supplementary. The same logic can be used to show that $\angle B$ and $\angle D$ are also supplementary.

8. In the diagram shown, *ABCD* is a square and ΔEGF is constructed so that *E* is the center of the square and *D* and *C* are midpoints of sides $\overline{EG}$ and $\overline{EF}$, respectively. What is the ratio of the areas of *ABCD* to *EGF*?

(See Reproducible 48, Square *ABCD* and ΔEGF.)

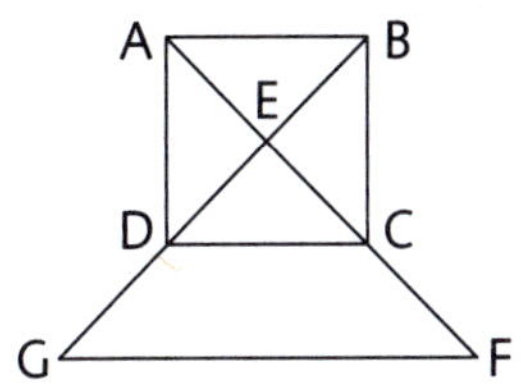

There are many ways for students to think about this problem. One way: since $\overline{DC}$ is a midsegment of *EGF*, it is parallel to and half the length of $\overline{GF}$. Therefore, the area of *EDC* is $\frac{1}{4}$ the area of *EGF* (because the side lengths of *EDC* are half the length of the corresponding sides of *EGF*). Since *EDC* is also $\frac{1}{4}$ of the area of square *ABCD*, the areas of *ABCD* and *EGF* must be equal. Therefore, the ratio is 1:1. Another way is to have students show how the three triangles *AED*, *AEB*, and *BEC* can tile trapezoid *GDCF*.

9. Which equations represent the quantities in the following diagram? Explain how you know.

(See Reproducible 49, Linear Angles.)

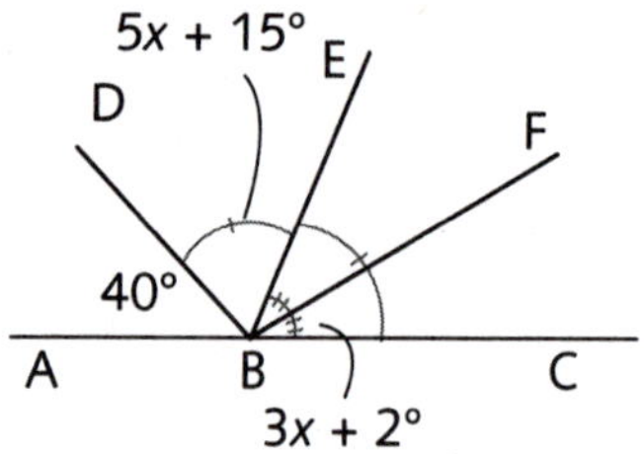

$5x + 15 = 2(3x + 2)$

$3x + 2 + 3x + 2 + 5x + 15 = 140$

$3x + 2 = \frac{1}{2}(5x + 15)$

$40 + 4(3x + 2) = 180$

These are all true. This question requires students to think more deeply about symbols and equations and interpret them as statements about relationships.

10. A square is subdivided into four parts by two perpendicular lines that intersect at the center of the square. What relationships can you find?

(See Reproducible 50, Square Cut by Two Perpendicular Lines.)

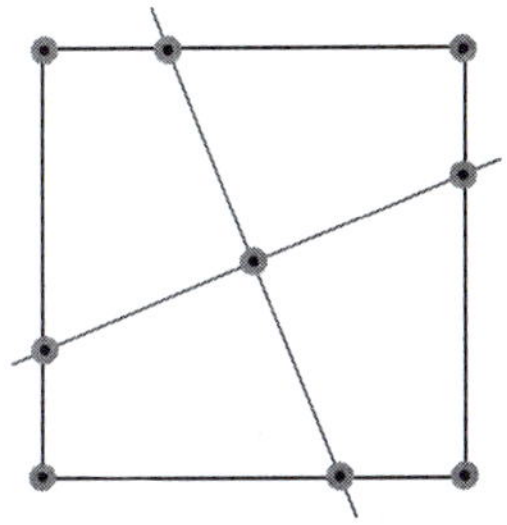

The four smaller quadrilaterals are all congruent to each other and have an area that is one-fourth of the entire square. To determine this, rely on the diagonals of the square (which are also perpendicular to each other and the bisectors) to identify eight small triangles. For example, in the following figure, the four shaded triangles are necessarily congruent to each other because vertical angles are congruent, the side lengths along the diagonals are congruent, and the angles at the vertices of the square are 45°. Similarly, the other four triangles are congruent to each other. Therefore, with these sets of congruent triangles, the quadrilaterals in the original figure can be shown to have corresponding angle measures and side lengths congruent.

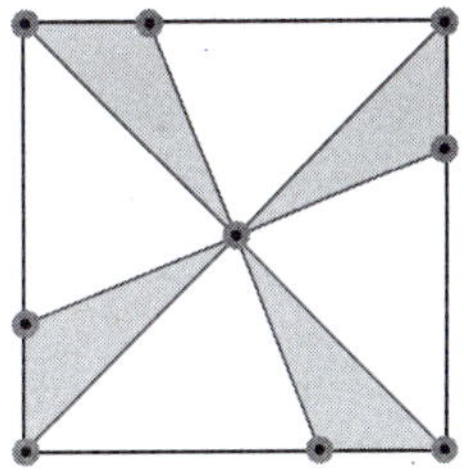

11. **A regular polygon has a central angle of 18°. How many sides must it have? What if the central angle is 40°? 62°?**

If the central angle is 18°, then it must have 360° ÷ 18 °= 20 sides. If the central angle is 40°, then it must have 360° ÷ 40° = 9 sides. However, the central angle of a regular polygon cannot be 62°, because 360° ÷ 62° is not an integer.

Right Triangles and Trigonometry

Experiences at This Level Will Help Students To

- understand, by similarity, side ratios in right triangles are properties of the angles in the triangle, leading to the definitions of the trigonometric ratios for acute angles
- explain and use the relationship between sine and cosine of complementary angles
- use trigonometric ratios and the Pythagorean theorem to solve problems involving triangles

Reproducibles are available in a downloadable, printable format. See page xx for directions about how to access them.

Materials

- Right Triangle *ADB* (Reproducible 51)
- Similar Right Triangles (Reproducible 52)
- Labeled Right Triangle (Reproducible 53)
- ΔABC with Altitude *h* (Reproducible 54)
- GQ's Triangle (Reproducible 55)

Good Questions and Teacher Notes (pages 194–204)

1. Yamile drew a right triangle *GQP* and found that sin $Q = \frac{5}{13}$. What can you determine about $\triangle GQP$?

Students should realize that there are an infinite number of triangles possible. They should also be able to conclude that the right angle is either *G* or *P*, that the measure of angle *Q* is approximately 22.62°, and that the other acute angle is approximately 67.38°. Challenge the students to create *GQP* with a hypotenuse of length 1; this seeds the concepts needed later when they study the unit circle. Ask students to create their own problems of this sort.

2. Driti was told that a right triangle included one side length of 15 cm and one angle that was 75°. What might the rest of the sides and angles be?

Ask students to discuss the different options (i.e., whether 15 cm is the measure of the adjacent leg, the opposite leg, or the hypotenuse). Have students discuss whether some choices are easier to solve than others and why. This allows them to confront the challenge of solving for the hypotenuse. Students can also discuss whether the complementary angle can be useful when using the tangent ratio.

3. AA and BB were considering the following right triangle with an altitude drawn from the vertex of the right triangle *ADB*. AA was given the lengths of exactly two segments in the figure and then she said that she could find any other side or angle measure in the figure. BB was also given the lengths of exactly two segments in the figure, but she concluded that she could not find the measures of any of the other segments or angles.

What information might each have been given?

(See Reproducible 51, Right Triangle *ADB*.)

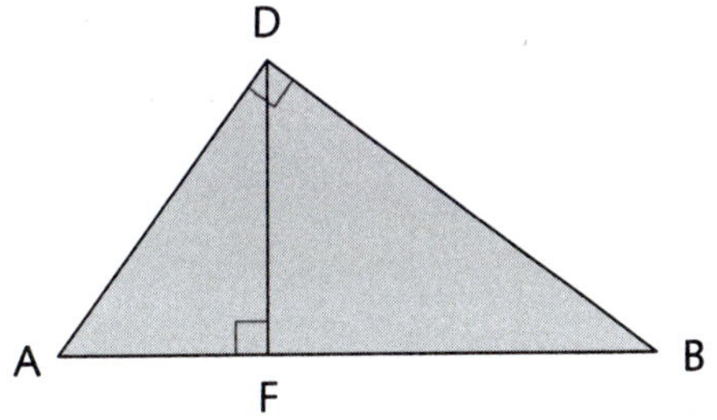

Students should realize that the figure contains three similar right triangles. For AA, a possible solution would be that she was given $AF = 3$ and $FB = 12$, in which case she could determine $m\angle A = m\angle FDB \approx 63.4°$ and $m\angle B = m\angle FDA \approx 25.6°$ and that $AD = 3\sqrt{5}$, $DF = 6$, and $DB = 6\sqrt{5}$. BB could have been given $AD = 3\sqrt{5}$ and $FB = 12$, which would not be enough information to find the other sides or angle measures.

4. In the following diagram, find the following missing information.

(See Reproducible 52, Similar Right Triangles.)

- If $AX = 5$, then $XB = ?$
- If $AC = 12$, then $YC = ?$
- Choose a reasonable value for the length of AD or AZ, then find the length of the other (AZ or AD).

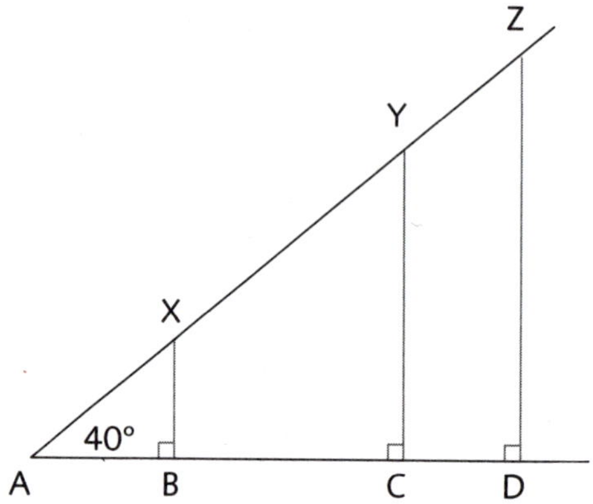

Here students are asked to solve two straightforward problems in finding that $XB = 5 \sin 40° \approx 3.2139$ and $YC = 12 \tan 40° \approx 10.0692$. For the third bullet point, ask which approach is easier, choosing AD or AZ? Students often struggle to solve for the hypotenuse AZ, so having students share insights and difficulties can be helpful.

5. Abasi was told that two sides of a right triangle were 6 and 12. What might the triangle look like and what could be the measures of the other side and the two acute angles?

Students may notice either the 30-60-90 right triangle where the missing side is $6\sqrt{3}$ *or* the triangle's hypotenuse is $6\sqrt{5}$ and the two acute angles would be 26.57° and 63.43°.

Ask students to create their own problems of this sort.

6. Create a real-world scenario in which there is a right triangle with a 25° angle and a 75-meter side. What are some questions that you could ask? What additional information would you need? Why?

A possible scenario could be a flagpole casts a 75-meter shadow when the angle of elevation of the sun is 25°. Assuming the ground is level, the height of the flagpole can be determined to be approximately 34.97 meters tall. Having students share their scenarios should expose the class to the different types of problems and approaches. You could vary this question by specifying what trigonometric ratio is desired.

7. Suppose that a triangle has side lengths 7, 10, and 12 units. What else can you determine about the triangle?

Students should be able to determine the perimeter, the area, and the angle measures of the triangle. You could then ask them to generalize solutions for sides a, b, and c. This could be used as a motivation for the Law of Sines and the Law of Cosines.

8. Santos says he has a triangle $\triangle GQP$ such that $\sin G = 2 \sin P$. What can you conclude about $\triangle GQP$? What might be the dimensions of $\triangle GQP$? What might be the angle measures of $\triangle GQP$?

Students should conclude that $PQ = 2GQ$. The following are possible diagrams for the problem. The diagram on the left is expected. Here students must realize that $\sin P \leq 0.5$. From there, there are infinitely many solutions depending on their chosen value for $\sin P$ and GQ.

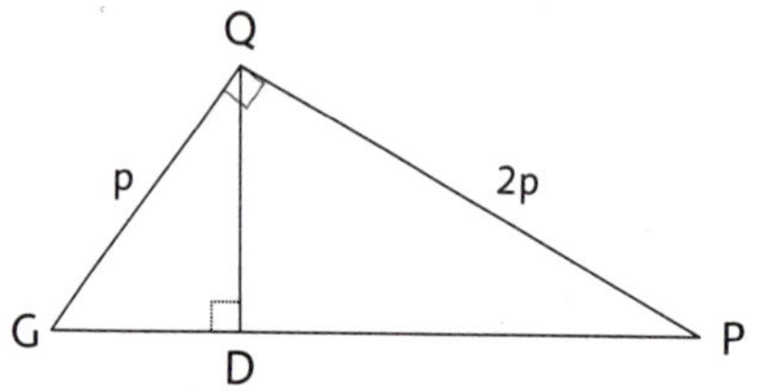

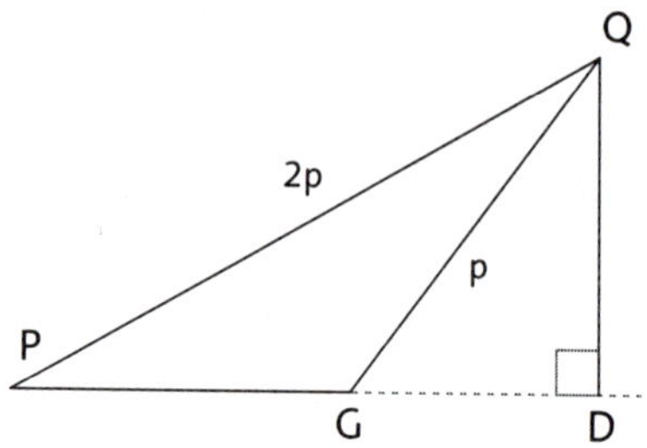

Special Case:

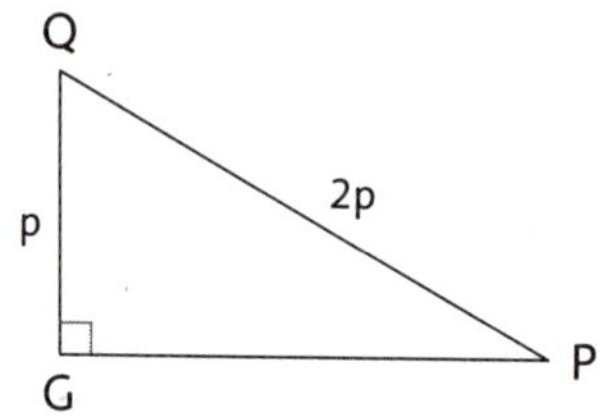

This question can be approached using right triangle trigonometry and then asked again after students have learned the Law of Sines or used to introduce the Law of Sines. You may want to scaffold the problem more by saying $\sin G = 0.6$ and $\sin P = 0.3$.

9. In right triangle *GQP* with right angle *Q*, fill in the blanks to create true statements about trigonometric ratios.

a. tan $P > 1$ when P is ______________________________.

b. $\sin^2 P + \cos^2 P$ is ____________________ when P is ____________________.

c. $\sin P > \cos P$ when P is ______________________________.

d. $\sin P \geq \cos Q$ when ____________________.

Students should notice that some solutions are conditional while others represent identities. Ask students to create additional statements to complete. Also, you can ask if the answers change when *GQP* is a non-right triangle, such as an obtuse triangle.

Note: Students may need help understanding that mathematicians prefer the notation $\sin^2 \theta$ to the equivalent $(\sin \theta)^2$.

10. Abe and Bea are talking about right triangles and trigonometry.

- Abe says that "in any right triangle, if you only know one of the sides and one of the acute angles, then you must use trigonometry to find the missing sides."
- Bea says she "mostly agrees, but if you know that one of the acute angles is 30°, 45°, or 60°, then you can find the missing sides without needing trigonometry."

Do you agree? Can you give examples that support Abe and Bea's reasoning?

This question is intended to have students practice right triangle trigonometry and to discover 30-60-90 and 45-45-90 triangles. Often students are introduced to special right triangles without addressing what is special about them.

11. **The three side lengths and the three angle measures of triangle *GQP* were each written on an index card. Alex selected three of the cards (at random), recorded the three components of *GQP*, and returned the cards. Britt and Caleb then each selected three of the cards at random as well and each recorded the three components they selected. Then each tried to use their trigonometric knowledge to solve for the remaining three components.**

If each student is correct, what information might each of them have selected?

- Alex says that she has solved for the remaining three components.
- Britt says that she has two different sets of solutions for the remaining three components.
- Caleb says that he cannot determine the remaining three components.

This question is intended to help students decide on appropriate strategies (Law of Cosines and Law of Sines) rather than scripting them. Encourage students to share their examples and reasoning. You could also scaffold the problem by providing the three components.

Alex either selected the three side lengths (Law of Cosines), two angles and any side (Law of Sines), one angle and the two adjacent sides (Law of Cosines), or one angle and an opposite side that is longer than the given adjacent side (Law of Sines or Cosines). Britt must have selected one angle and an opposite side that is shorter than the given adjacent side (Law of Sines or Cosines). Caleb selected the three angles and thus only knows the shape.

12. Xantheus and Chi are investigating right triangles and trigonometry. Xantheus thinks that when you draw a right triangle in a particular way, the tangent of an angle is the slope of the hypotenuse. Chi says that she keeps getting the reciprocal of the slope. In what way is Xantheus drawing his right triangle? Why is Chi getting the reciprocal?

This question is intended to relate the rise/run to the opposite/adjacent, which introduces the standard position that will be used later when extending the definitions of trigonometry to the unit circle and circular functions. So the diagram should resemble the following:

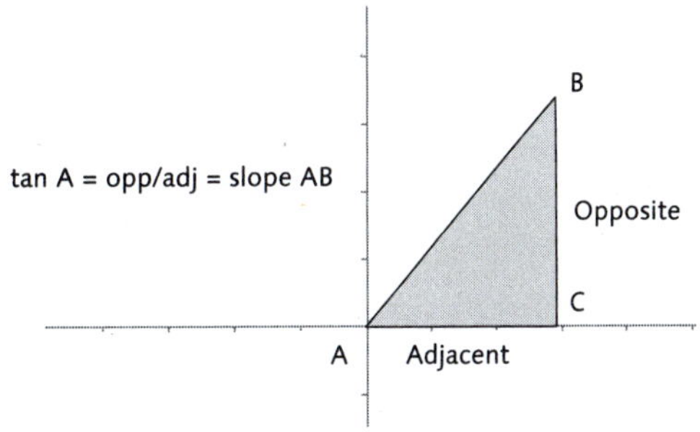

You can extend this question beyond right triangles to include obtuse triangles where the tangent is negative (as is the slope of the hypotenuse of the related right triangle).

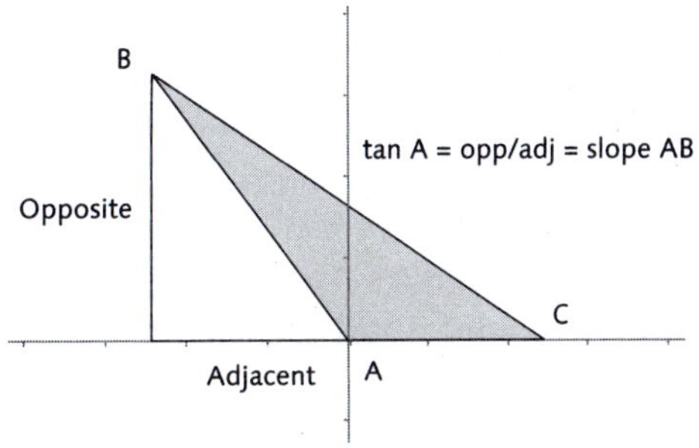

13. For ΔGQP, we know that $PG = 1$, $\sin P = \frac{\sqrt{3}}{2}$, and $\cos P = -\frac{1}{2}$. What might ΔGQP look like?

This question introduces students to the world of obtuse triangles and the fact that the cosine of an obtuse angle is negative, but its sine is still positive. This could be used either after you have introduced students to the trigonometry of obtuse angles or as a start of that discussion with some scaffolding.

14. Write as many true statements as you can for this diagram.

(See Reproducible 53, Labeled Right Triangle.)

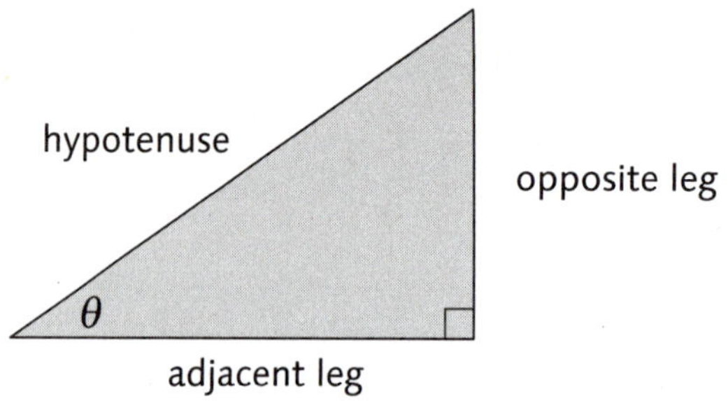

Here you would expect that $(\text{adjacent leg})^2 + (\text{opposite leg})^2 = (\text{hypotenuse})^2$, $\sin\theta = \frac{\text{opp leg}}{\text{hyp}}$, and $\cos\theta = \frac{\text{adj leg}}{\text{hyp}}$ are among the true statements. Then you can lead students to divide the Pythagorean equation by $(\text{hypotenuse})^2$ to yield the important result $\sin^2\theta + \cos^2\theta = 1$ for all θ.

Note: Students may need help understanding that mathematicians prefer the notation $\sin^2\theta$ to the equivalent $(\sin\theta)^2$.

15. **Given this figure, write as many true statements as possible using your knowledge of triangles.**

Suppose that you know the values of a, b, and $\angle C$. Can you use some of your true statements to find the value for c?

(See Reproducible 54, ΔABC with Altitude h.)

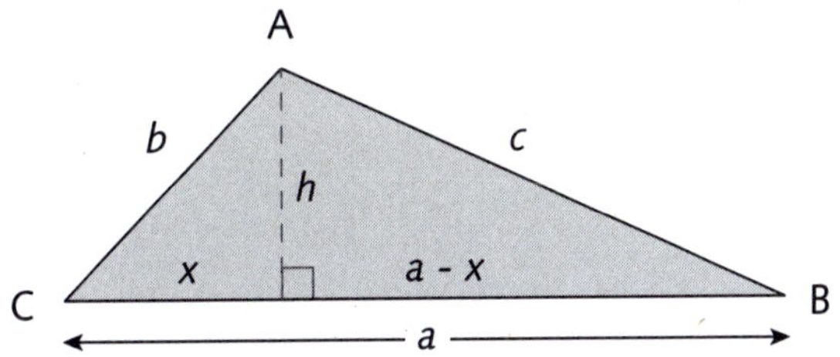

This question is useful when deriving the Law of Cosines. If students have noted that $\cos C = \frac{x}{b}$, $\sin C = \frac{h}{b}$, and $c^2 = h^2 + (a - x)^2$, they can use substitution and the fact that $x = b \cos C$ and $h = b \sin C$. If the students have not yet discovered that $\sin^2\theta + \cos^2\theta = 1$ for all θ, then you may want to ask the previous question first.

16. GQ says that she can find the area of a triangle given two sides and the measure of their included angle. Using the following diagram for reference, what expressions might GQ write for the area of the triangle?

(See Reproducible 55, GQ's Triangle.)

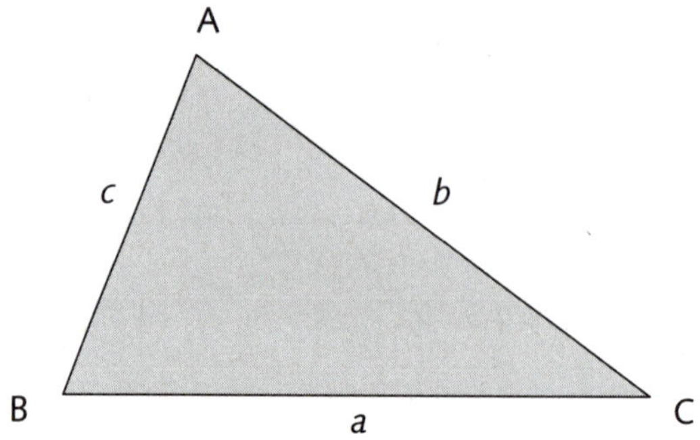

Students should be expected to have $\text{Area}_{\Delta ABC} = \frac{1}{2}ab \sin C = \frac{1}{2}ac \sin B = \frac{1}{2}bc \sin A$ among their expressions. Then, asking students to multiply each by 2 and dividing each by *abc* should yield the Law of Sines.

Students can also use graphing software (or ruler, protractor, pencil, and paper) to create triangles, measure the sides and angles and confirm that the Law of Sines is valid. Perhaps doing this first will lead students to the derivation.

17. Create a real-world scenario that could be solved using the Law of Cosines. Be prepared to share your problem and your solution.

This question can be adapted to require the Law of Sines as needed. Ask students to discuss whether there are alternate approaches to solve the problem.

CHAPTER 8

Good Questions for Statistics and Probability

Decisions or predictions are often based on data—numbers in context. These decisions or predictions would be easy if the data always sent a clear message, but the message is often obscured by variability. Statistics provides tools for describing variability in data and for making informed decisions that take it into account. To these ends, students need a comprehensive experience in learning the techniques of exploratory data analysis to make sense of data; applying and analyzing the techniques for producing and gathering data; learning that statistical models provide a structured framework for the analysis; and understanding the role of randomization and probability in making decisions.

Interpreting Categorical and Quantitative Data

Experiences at This Level Will Help Students To

- summarize categorical data for two categories in two-way frequency tables and interpret relative frequencies in the context of the data (including joint, marginal, and conditional relative frequencies) and recognize possible associations and trends
- represent data of two quantitative variables on a scatterplot and describe how the variables are related
- use linear, quadratic, exponential, or other functions fitted to data to solve problems in the context of the data
- informally assess the fit of a function by plotting and analyzing residuals
- fit a linear function for a scatterplot that suggests a linear association
- interpret the slope (rate of change) and the intercept (constant term) of a linear model in the context of the data
- compute (using technology) and interpret the correlation coefficient of a linear fit
- distinguish between correlation and causation

Reproducibles are available in a downloadable, printable format. See page xx for directions about how to access them.

MATERIALS

- Data Display of Sugar Content of Adult versus Children's Cereal (Reproducible 56)
- Plot of Sugar in Cereal by Display Shelf (Reproducible 57)
- Turtle Weights Dot Plot (Reproducible 58)
- Plot of Baseball Salaries in 2017 (Reproducible 59)
- Data Displays of Cereal Nutrition (Reproducible 60)
- Table of Temperature Statistics by Season (Reproducible 61)
- Sets of Die-Rolling Results (Reproducible 62)
- Nine Point Scatterplot (Reproducible 63)
- Scatterplots A–I (Reproducible 64)
- Z. Oologist's Dot Plots (Reproducible 65)
- Z. Oologist's Scatterplot (Reproducible 66)
- Scatterplot and Linear Model (Reproducible 67)
- Scatterplot with Data (Reproducible 68)
- Earnings of Domestic and Worldwide Movies (Reproducible 69)

GOOD QUESTIONS AND TEACHER NOTES (PAGES 208–251)

1. **Abby is interested in the Summer Olympics. She gathers data on the 2016 Summer Olympics and learns that 85 different countries won at least 1 medal and the distribution of medals won by those medal-winning countries is skewed right with 9 high outliers. Create an appropriate plot that could represent this data.**

Students have the option to draw a boxplot or a histogram (or perhaps a dot plot). Two possible displays are:

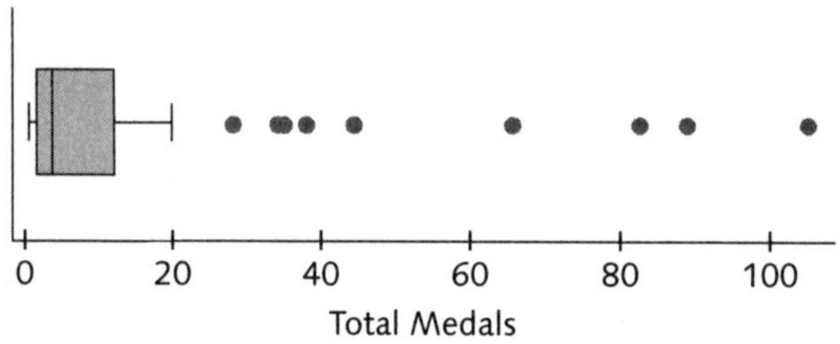

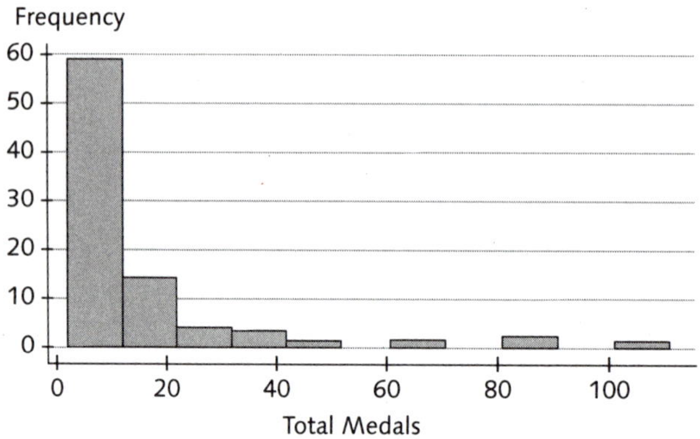

As an extension, provide students with additional details and ask how they might modify their answers, (e.g., Min = 1, Q1 = 2, median = 4, Q3 = 12, Max = 104). Ask students to estimate the mean number of medals and interpret what that means and why it may be misleading to use that value.

2. **Abby is interested in the Summer Olympics. She gathers data on the 2016 Summer Olympics and learns that 85 different countries won at least 1 medal; the mean number of medals won per country (for those winning at least one medal) is 11.32 medals and the median is 4 medals. Explain why these two values are so different.**

Based on the information, ask students to determine the number of medals awarded at this Olympics.

Ask students if the mean number of medals per country participating (including those who did not win any medal) would be different. What about the median medals per participating country?

It is important that students understand that the mean still provides useful information even when it is not the best descriptor of the "center" or "typical" value.

3. John and Otello have each created plots of single (different) quantitative variables.

- John looks at his plots and says, "The best descriptors of center and spread would be the median and interquartile range."
- Otello looks at his plots and says, "I can use the mean and standard deviation to describe my data's center and spread."

What might their plots look like?

Students should identify skewed distributions and/or distributions containing outliers for John, while plots for Otello should be symmetric. One possible answer is shown on the next page.

Have students discuss the type of plot they create (histogram, dot plot, or boxplot) and the rationale for their choice.

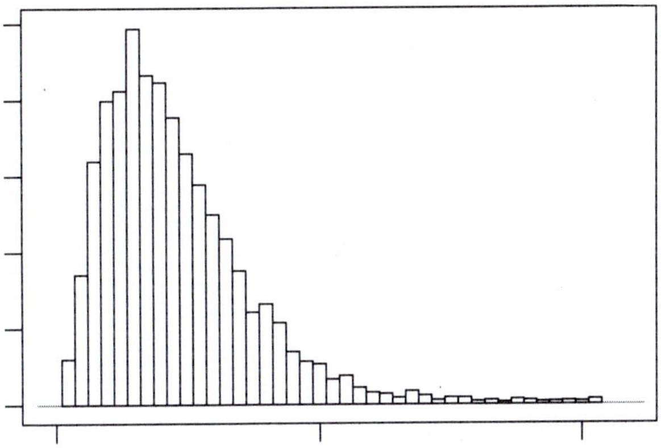

John's plot

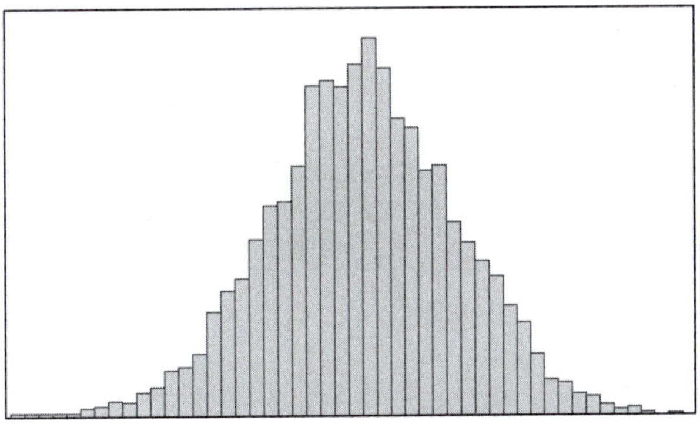

Otello's plot

4. A father and son are looking at nutritional data on cereals. They decide to focus on the grams of sugar in a serving for cereals grouped by *A* (adult) and *C* (children). Use the following data display and statistics to compare the sugar contents for these two types of cereal.

(See Reproducible 56, Data Display of Sugar Content of Adult versus Children's Cereal.)

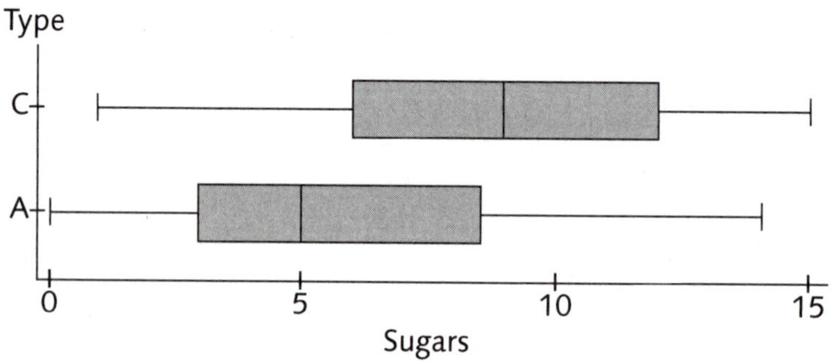

Type	*n*	Mean	Standard Deviation	Minimum	Q1	Median	Q3	Maximum
A	44	5.6363636	4.2266646	0	3	5	8.5	14
C	33	8.6666667	4.1281554	1	6	9	12	15

Here, the distributions are both roughly symmetry, so students can use any of the statistics to support their distributions. Students should discuss why using the medians will demonstrate a more significant difference in sugar contents.

Students should use the quartiles to make specific statements, for example: More than 75 percent of the children's cereals have more sugar than the median sugar content in adult cereals.

5. Students interested in the marketing of cereals decided to create a plot of the grams of sugar per serving grouped by the shelf on which the cereal was displayed at the grocery store.

How would you compare the cereals on the three shelves?

Why do you think the cereals are displayed this way?

(See Reproducible 57, Plot of Sugar in Cereal by Display Shelf.)

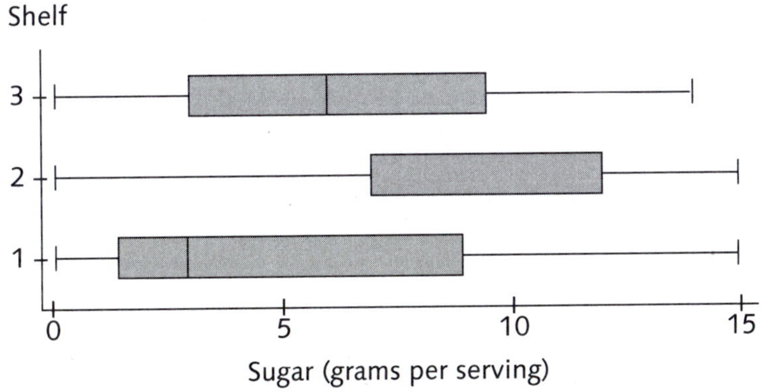

Be sure to discuss why the boxplot for Shelf 2 has no apparent median drawn. Do students think the median is also Q1 or Q3? Why? (For Shelf 2, the median = Q3 = 12 grams of sugar per serving.)

Students should notice that while Shelf 3 has a symmetric distribution, Shelf 1 is skewed right and (it turns out) Shelf 2 is skewed left. Ask what effect comparing the mean sugar contents on shelves 1 and 2 would reveal.

Ask students to interpret differences in shape, center, and spread in the context of the data sets.

6. Beatrice produced the following dot plot to examine the weights of a particular type of sea turtle bred in captivity. She also calculated that the mean weight is 250.4 pounds but she is not sure whether that is appropriate. She also wonders if the median weight of 246 pounds is appropriate. What could be the source of her concerns?

(See Reproducible 58, Turtle Weights Dot Plot.)

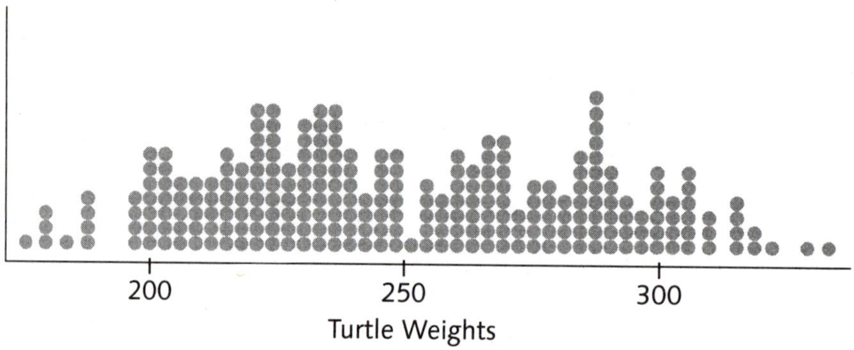

Students should notice that the distribution of weights is bimodal, which indicates that the data likely has two categories of turtle. Ask students to speculate as to the difference. For example, since they are bred in captivity, the two groups could be adults and offspring, or students may suggest that male and female turtles differ in weight.

Ask the students to create a possible dot plot by group (by whatever factor they determine is contributing to the bimodal nature).

7. It has been said that "pitching is the most important component to winning in baseball." The following plot shows the salaries (*USA Today*, 2017) for pitchers (*P*), fielders (*F*), and designated hitters (*DH*). Does this information support that claim? Explain your reasoning.

(See Reproducible 59, Plot of Baseball Salaries in 2017.)

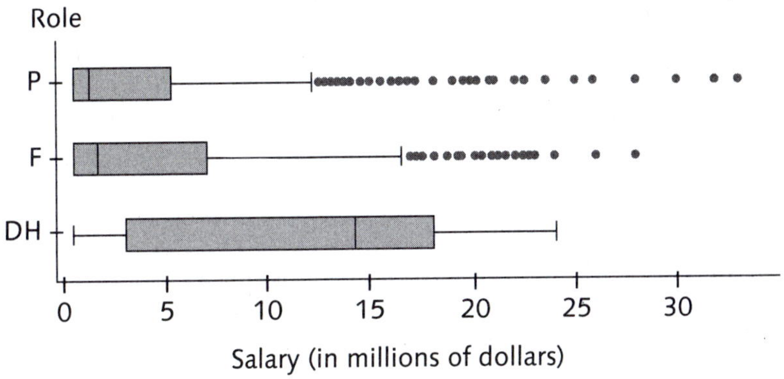

This question has no clear answer, since pitchers had the lowest median salary of the three, while also garnering the highest salaries overall. For example, students here could report that the median salary for fielders ($2,000,000) is slightly higher than the median salary for pitchers ($1,500,000), and that the median salary for designated hitters ($14,000,000) is higher than all but the highest paid pitchers (all outliers), and so that is an argument against the claim. However, the highest paid players were all pitchers, which students can say is some evidence in favor of the claim.

8. A group of students were interested in the nutrition data for cereals. They compiled data for 77 cereals. The displays on the next page were created to investigate the sugar content of those cereals. Which displays are most helpful? Why? What are some possible conclusions? Note: The mean and median are plotted on each graph.

(See Reproducible 60, Data Displays of Cereal Nutrition.)

Students should share their choices and rationale. Perhaps gather data as to their choices to demonstrate how data can come from a variety of sources. For example, they might notice that the dot plot and histogram of Percent of Calories from Sugar seem to show two clusters, one between 0 and 15 percent and another between 20 and 50 percent, so while the distributions of Grams of Sugar per Serving and Percent of Calories from Sugar are fairly symmetric and centered at 7 grams or 25 percent (a picture easily seen in the boxplots), there are really two different types of cereals that require the more detailed nature of the histogram and the fully detailed nature of the dot plots.

You might also ask students what other displays they would want to see. What other data might they collect?

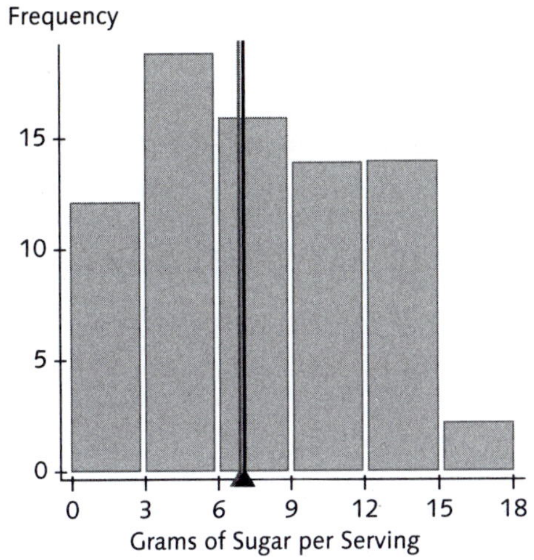

Histogram 1

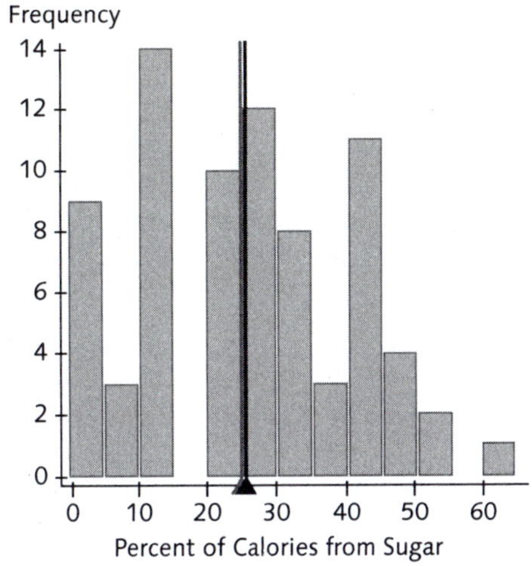

Histogram 2

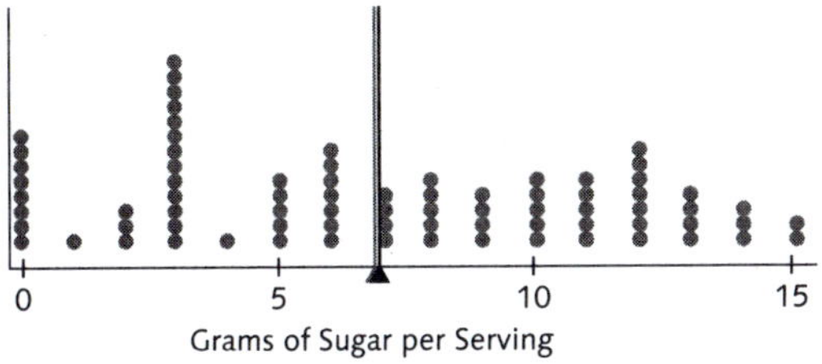

Dot plot 1

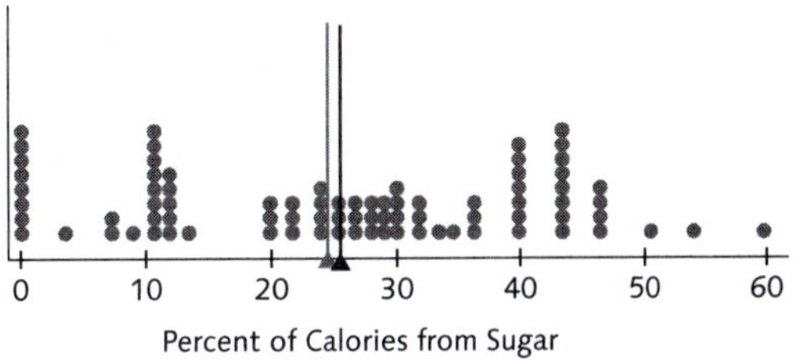

Dot plot 2

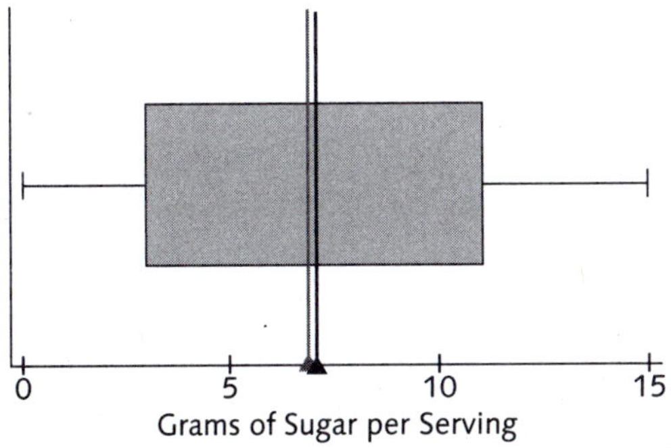

Boxplot 1

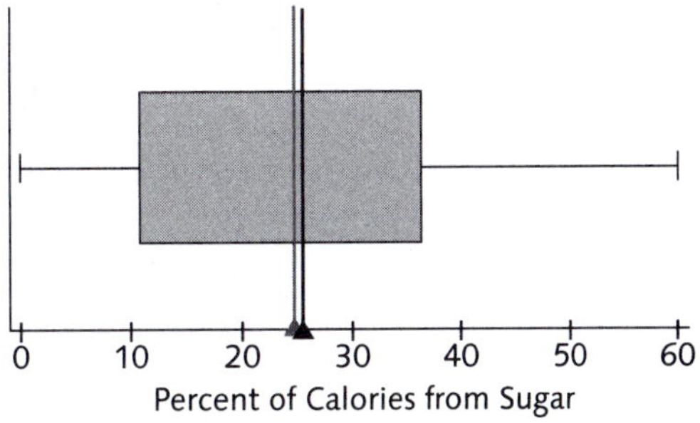

Boxplot 2

9. In professional sports there are often disputes about player salaries between the owners and the players' union.

In 2017, the owners of 30 teams and the players' union (representing 868 players) disagreed about the "average" player's salary. The owners claimed that the "average player" earned $4.7 million while the players' union claimed that the "average player" earned $1.6 million (*USA Today*, 2017).

Meanwhile, you discovered that the most common salary was $535,000. Is it possible that all three statements are correct? Explain.

Students should discuss the vague term *average* versus the statistically specific terms *mean*, *median*, and *mode*. Ask students to use the information provided to figure out how much money was spent on player salaries in 2017.

As an extension, ask students to look for other examples where mean, median, and mode can tell very different stories about the same data set.

10. For each descriptor of shape, state examples of quantitative variables whose distributions are likely to have that descriptor.

a. Uniform

b. Mound-shaped and symmetric

c. Bimodal or multimodal

d. Skewed right

e. Skewed left

This question encourages students to be specific in their language to include the variable of interest as well as the population from which it is measured. Possible examples include the following: The number of dots showing on a large sample of rolls of a fair six-sided die (number cube) should be uniform. The distribution of test scores on an "easy" statistics test would probably be skewed to the left, while the scores on a pretest of new material would be more skewed right or symmetric.

The question could be flipped to focus on quantitative variables. For example, have students state a quantitative variable, and then have the class discuss what the shape of its distribution should be.

11. In a particular forest, the average daily temperature (in °C) for each day of the year is recorded and the data are grouped by season. Use the summary statistics provided to compare and contrast the distribution of average daily temperature for each season.

(See Reproducible 61, Table of Temperature Statistics by Season.)

Summary Statistics for Average Temperature (°C):

Season	Mean (°C)	Standard Deviation (°C)	Median (°C)	IQR (°C)
Fall	3.34	9.29	5.83	15.62
Spring	12.87	6.31	13.85	9.35
Summer	19.60	3.22	20.1	4.80
Winter	-4.32	6.30	-3.67	9.11

Here, students should be noticing the relationship between the mean and median temperatures. As expected, the temperatures are warmest in summer, coolest in winter. Students should also notice that the temperatures in the fall tend to be more skewed left since the mean temperature is much lower, 3.34 versus the median temperature of 5.83. Also, students should notice that the fall has the most variability in temperature with an IQR of 15.62 and that the summer has the least variability (IQR 4.8) and most symmetry since the mean and median temperatures are similar at 19.60 and 20.10, respectively.

12. **For each of the following, create a set of nine numbers with the stated characteristics or explain why it is not possible:**

a. More than half of the numbers are greater than the mean.

b. More than half of the numbers are greater than the median.

c. The mean is greater than the median.

d. Doubling each data point would not change the mean value.

e. The interquartile range is greater than the standard deviation.

f. The interquartile range is 0.

You may choose to adapt this question by only listing one stated characteristic at a time.

Students need to work with small data sets to get a better sense of the interplay between various statistics. Using a spreadsheet can help them test hypotheses more quickly once they have done some by paper and pencil.

Students can also be challenged to create sets of nine data points that satisfy more than one of these characteristics.

13. **Mr. Fisher has two statistics classes. On the final exam he noticed that three-fourths of the students in one of the classes did as well or better than the typical student in the other class. Sketch two parallel boxplots to satisfy this observation. Based on your sketch, what are some other possible observations?**

A possible extension for this question is to add that the two classes had the same median score.

14. Carlos collects data on the eruptions of Old Faithful at Yellowstone National Park. He wants to model the eruption durations using a normal distribution, so he finds that the mean eruption duration was 3.58 minutes with a standard deviation of 1.08 minutes.

After looking at a histogram of the eruption durations, Carlos realizes that a normal model would be inappropriate. What might the histogram have looked like? What distribution was he hoping to see?

You could extend this question, since the eruption durations are bimodal, by asking students to find data on the eruption times to confirm the bimodal nature of the eruptions. Also, students could extend the question to find data on the times between eruptions, which is also bimodal.

15. Mira is training for a cycling race and learns that her time on the course is two interquartile ranges (IQRs) below the median time for the competition. Construct a scenario for which Mira's score would not be considered an outlier.

Clarify that students should use the established definition of an outlier being more than 1.5 IQRs away from its respective quartile. Ask students if the normal model could be appropriate. (It would not, since the cutoff for outliers in a normal model is 2.02 standard deviations from the median and Mira's time would be 2.7 standard deviations from the median.)

16. Ms. Oskooi has two statistics classes. She asks the students in each class to complete a Sudoku puzzle and records the time it takes for each student to finish. She notices that the average times are essentially the same for each class, but there is much greater variability in his afternoon section. Sketch two boxplots to satisfy Ms. Oskooi's description.

A possible solution could look as follows where the means were similar, but both the boxes and whiskers for the afternoon section are wider:

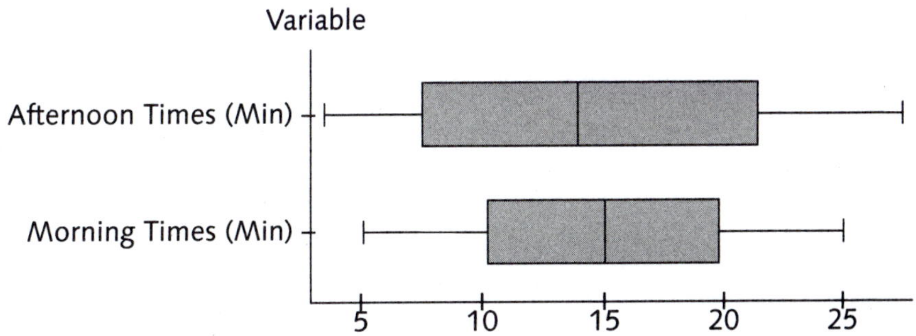

Students should realize that the IQR (box width) is the more important consideration than the whiskers since the boxplot does not display the distribution within the extremes unless the values are outliers.

17. Hermione and Luna are studying the medals won by the various countries at the 2012 Summer Olympics.

They notice that three-quarters of the 85 medal-winning countries win 12 medals or fewer and there are 10 countries whose number of medals won are considered outliers, with the United States winning the most, 104.

Hermione draws a boxplot of the number of medals won by each medal-winning country. Luna creates a bar graph of the 10 countries whose medal winnings were outliers. What might the two graphs look like?

Students often have a difficult time shifting between categorical variables and quantitative ones, and so the following are possible representations:

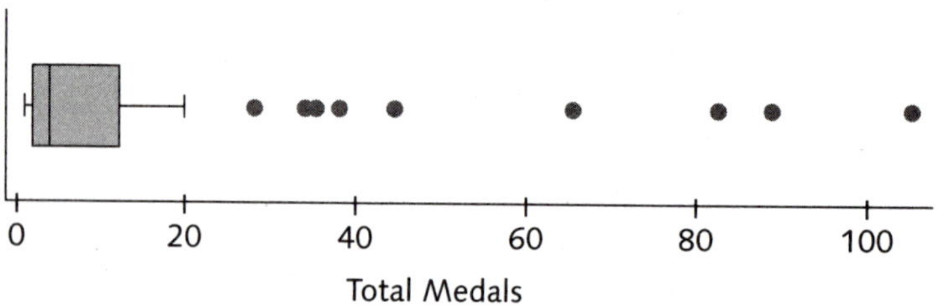

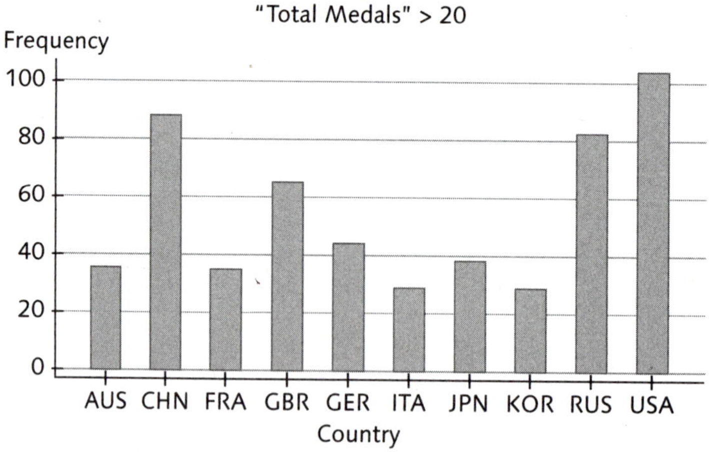

Ask students to look at the following graph and explain what the tall bars on the left represent. Often, students will confuse the height of the bar with the number of medals won rather than the number of countries in that band of total medals won.

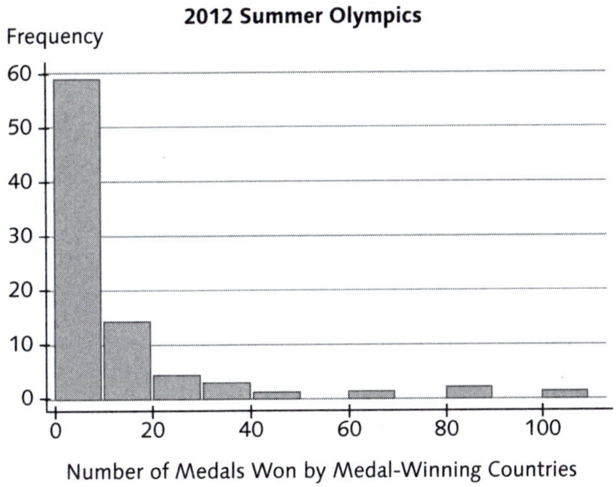

18. Alex works in quality control at the Fair Chip Company, and he has discovered that the weights of the contents in Family Size (13.5 ounces on the label) bags of potato chips follow a normal distribution with a standard deviation of 0.25 ounces. Alex needs to decide the actual target amount of chips to put in each bag.

What would you suggest and why?

Ask students if it is acceptable that some bags actually contain less that the stated 13.5 ounces. Given the standard deviation of 0.25 ounces, what target amount would be acceptable? (For example, if the target amount is 13.75 ounces then roughly 16 percent of the bags would have less than the stated 13.5 ounces on the label.)

19. Gary is training for a cycling race and knows that his mean time for the course is two standard deviations below the mean of all the competitors.

If those times can be modeled using the normal model, about what percent of the competition is faster than Gary?

How would your answer change if the distribution of times were skewed right? Skewed left?

This question encourages students to apply the empirical rule. If you want to assess their use of the model further, change the number of standard deviations to a nonintegral value.

If the times follow a normal model, then roughly 2.5 percent of the racers are faster. If skewed right, then fewer than 2.5 percent are faster, whereas if skewed left would imply that more than 2.5 percent are faster.

20. Suppose the distribution of a large set of quantitative data can be modeled by the normal model.

What proportion of the data will be considered outliers?

Would it still be appropriate to use the mean and standard deviation to describe this distribution's center and spread?

Since there is a rule of thumb that outliers are typically considered to be any value that is either 1.5 IQRs above Q3 (z-score = 0.67) or below Q1 (z-score = -0.67), we can apply that to show that the IQR is 1.34 standard deviations wide and so outliers would be any point more than 2.01 standard deviations away from the mean. We would expect roughly 5 percent of the values to be considered outliers by this rule. Yes, the mean and standard deviation are still appropriate since the outliers should be symmetrically distributed both above and below the mean.

21. Maya was shopping for a Bluetooth speaker online. She read the ratings (1–5 stars) for one model and found that it received twice as many 3-star ratings as 2-star ratings; 10 percent more people gave a 4-star rating than a 3-star rating; and 70 percent gave it a 5-star rating. What would a bar graph of this speaker's ratings look like?

This question encourages students to reflect on the idea that there are two ways to compare: *difference* and *relative ratios*. A possible solution is shown.

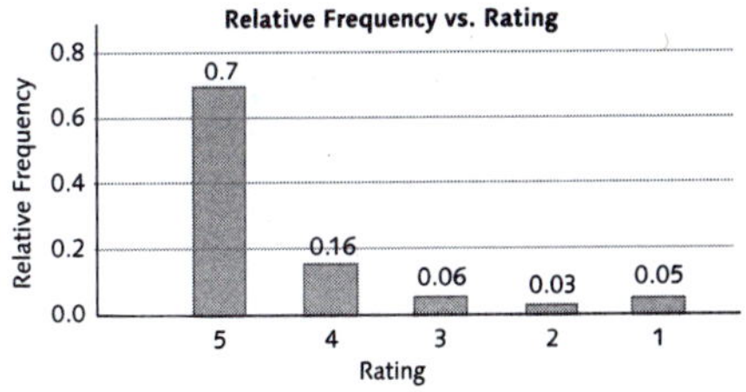

Ask students if they would buy this speaker (if price was not a factor). Why or why not?

22. Bill and Melinda are trying to choose between two new restaurants—Alba and Brava. They look at ratings on a popular restaurant review site and find that both restaurants have an average rating of 3.5 stars (out of 5). Bill suggests Alba because many people rave about the food. Melinda suggests Brava because it seems the best chance for a decent meal. What might bar plots of the ratings be showing?

A possible solution might look like the following, where they both have an expected rating of 3.5; but while Alba shows many 4 and 5 ratings, there are a significant number of 1 ratings, whereas Brava shows virtually all ratings of 3 or above. The focus is on how the ratings are distributed around the typical value.

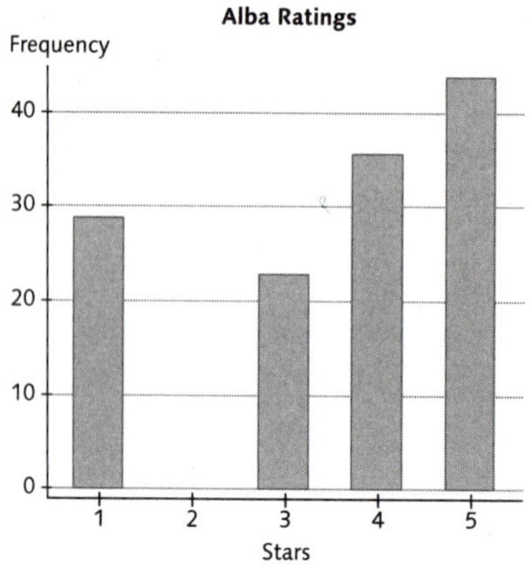

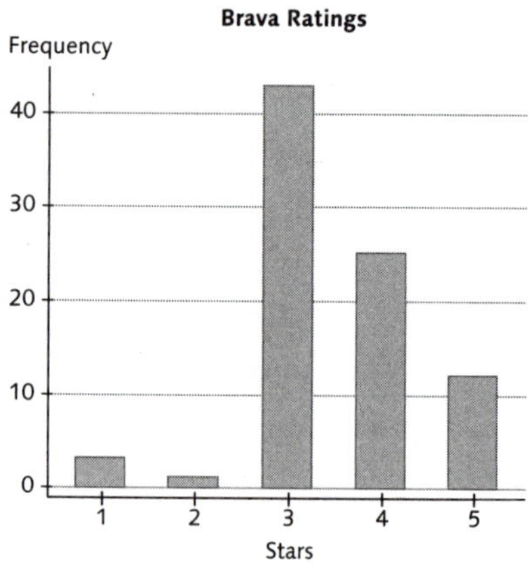

23. Following are three hypothetical results of rolling a six-sided die to test whether it is a fair die.

Which set would be the least proof that the die is unfair? Most proof? Explain.

Can you create a single numerical measurement (statistic) that would quantify your reasoning for these choices?

(See Reproducible 62, Sets of Die-Rolling Results.)

Set I	
Pips	Frequency
1	13
2	7
3	9
4	12
5	13
6	6
Total	60

Set II	
Pips	Frequency
1	130
2	70
3	90
4	120
5	130
6	60
Total	600

Set III	
Pips	Frequency
1	103
2	97
3	99
4	102
5	103
6	96
Total	600

Here, Set II provides the strongest evidence that the die is unfair, and Set III provides the least.

Here we are setting up the concept of the chi-square test statistic, which is expressed by the following formula and which contains the common use of squaring deviations to quantify the spread of a data set, similar to standard deviation for quantitative variables.

$$\chi^2 = \sum \frac{(\text{Observed} - \text{Expected})^2}{\text{Expected}}$$

24. M&M candies come in six traditional colors (blue, brown, green, orange, red, and yellow). Max thinks that the colors are equally distributed. Minny believes that brown is the most common and that green, blue, and yellow are rarer. What color distributions are Max and Minny expecting to see?

You could get bags of M&M candies and have students discuss whether or not it confirms or refutes either Max's or Minny's beliefs.

25. Joe Student is researching colleges and comparing two universities, Good Questions University (GQU) and Good Questions State University (GQSU). Joe has data from the previous year's applicants that records which college an applicant applied to, an applicant's gender, and the acceptance decision. Both schools have three colleges to which an applicant can apply (Business, Engineering, and Liberal Arts), and each university typically gets 10,000 applicants to fill 2,000 slots.

Joe notices that applicants' acceptance decision at GQU is associated with the school to which they applied; however, that is not the case at GQSU.

Create a two-way table for each university's applicants that could support these observations.

Joe notices that while female applicants have a higher proportion of acceptance in each college at GQU, male applicants actually have a higher proportion of acceptance at the university.

Create data that supports this observation at GQU. If the college applied to is *not* associated with the acceptance decision at GQSU, is it possible to observe the paradoxical situation Joe found at GQU? Why or why not?

Possible Data for Joe's Observation for Admission at GQU versus GQSU

	GQU Accepted	GQU Rejected	GQU Total	Acceptance Rate	GQSU Accepted	GQSU Rejected	GQSU Total	Acceptance Rate
Engineering	200	900	1,000	.222	400	1,600	2,000	.20
Business	600	3,200	6,000	.10	800	3,200	3,000	.20
Liberal Arts	1,200	4,000	3,000	.40	1,000	4,000	5,000	.20
Totals	2,000	8,000	1,000	.20	2,000	8,000	10,000	.20

This task asks students to create a situation known as Simpson's paradox, which can occur when data is aggregated (here, when ignoring the *college applied to* variable). Following is a table showing one possible solution. The key here is that the College of Liberal Arts has a much higher proportion of acceptances than the other two colleges, and the male applicants were much more likely to apply to that college, which brings the male overall acceptance rate up, while the female applicants tended to apply to the other two schools, which brought their overall acceptance rate down.

	Male Accepted	Male Rejected	Female Accepted	Female Rejected	Total Accepted	Total Rejected	Total
Engineering	50	450	300	1,700	350	2,150	2,500
Business	50	450	300	1,700	350	2,150	2,500
Liberal Arts	1,000	3,000	300	700	1,300	3,700	5,000
Totals	1,100	3,900	900	4,100	2,000	8,000	10,000

26. Lian and Belinda are creating scatterplots. Lian says that her scatterplot shows a moderately strong negative linear association; however, when one point is removed, she says there doesn't appear to be any linear association anymore. What might their original scatterplots look like?

One possible answer is shown here:

The plot shows a correlation of -0.883, which implies a strong negative linear association.

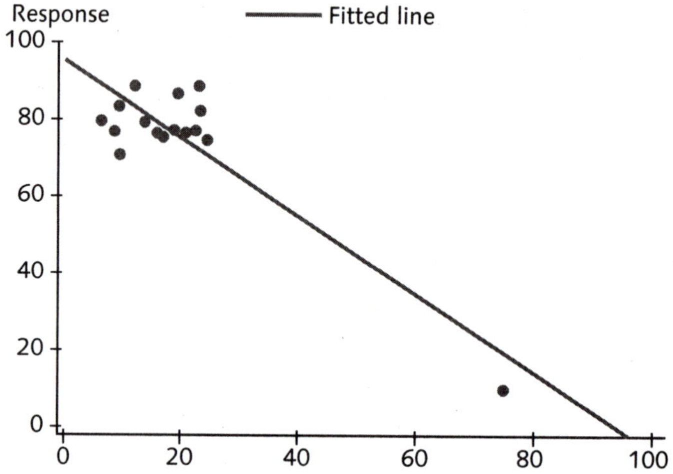

Simple linear regression results:

Dependent Variable: *y1*

Independent Variable: *x1*

y1 = 95.275845 – 1.0124531 *x1*

Sample size: 16

R (correlation coefficient) = -0.8827596

R-sq = 0.77926452

Estimate of error standard deviation: 8.7921729

However, when the *influential point* is removed, we see that the correlation is 0.088, showing virtually no association. Discuss the importance of students' looking at scatterplots along with statistical values to detect outliers and influential points.

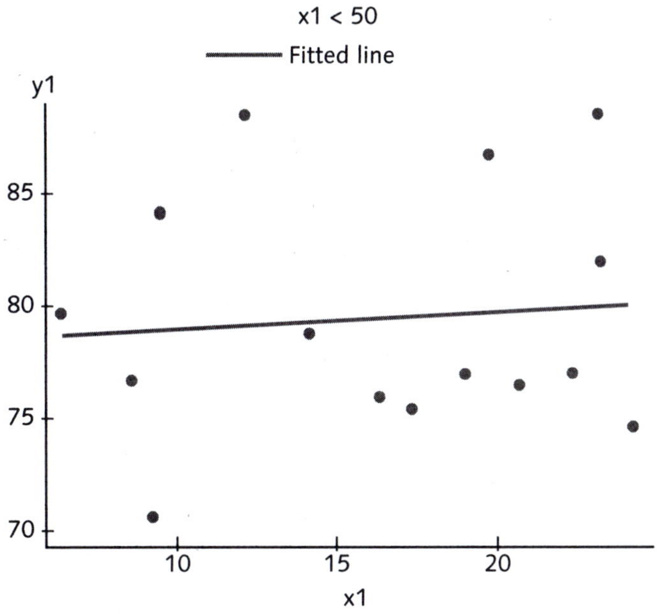

Simple linear regression results:

Where: *x1* < 50

Dependent Variable: *y1*

Independent Variable: *x1*

y1 = 78.068008 + 0.078081594 *x1*

Sample size: 15

R (correlation coefficient) = 0.088092974

R-sq = 0.007760372

Estimate of error standard deviation: 5.4887529

27. Suppose that you are given the following scatterplot. If you could add another point (with integer coordinates) to the scatterplot within the window shown, what point would you add in order to accomplish each stated goal? If there is no such point, explain why.

(See Reproducible 63, Nine Point Scatterplot.)

Solve for the following:

a. Make the correlation as high as possible.

b. Make the correlation close to 0.

c. Increase the correlation as little as possible, but some.

d. Leave the correlation unchanged.

e. Decrease the correlation as little as possible, but some.

f. Make the correlation negative.

g. Make the correlation as close to -1 as possible.

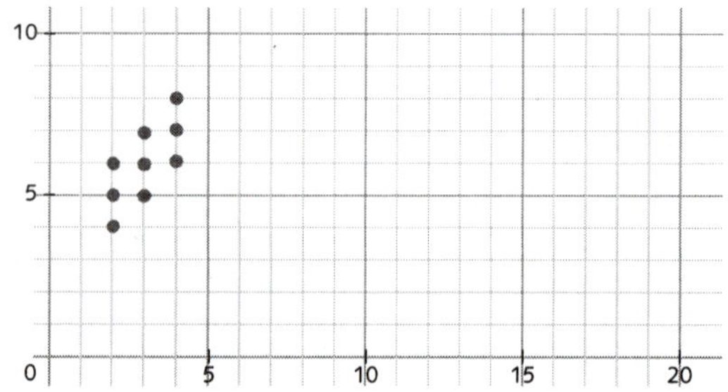

Students are encouraged to verify or correct their answers using technology like Desmos, spreadsheets, or graphing calculators.

28. Suppose you are given the following two sets of numbers:

Explanatory Set = {4, 7, 16, 22, 34}

Response Set = {4, 12, 16, 22, 24}

You are asked to pair each member of the explanatory set with a unique member of the response set. Use an app or graphing software to create pairings for each of the following goals:

a. the strongest possible positive (linear) association

b. the strongest possible negative (linear) association

c. a negative (linear) association that is not the strongest possible

d. a positive (linear) association that is not the strongest possible

e. the weakest possible (linear) association

This question allows students to form their own understanding of the strength and direction of an association between two quantitative variables.

29. Consider the scatterplots on the next page.

(See Reproducible 64, Scatterplots A–I.)

a. Which associations appear linear?

b. Which associations are the strongest?

c. Arrange the correlations from smallest to largest.

d. Notice that each plot has horizontal and vertical lines to create four quadrants.

What do you notice about the distribution of the points to these quadrants? What do you think the lines represent?

This question is designed to distinguish between associations between variables and the specific concept of correlation that deals only with linear associations. While Plots C and H both show strong associations, neither is linear. Plot C has a correlation close to 0 while Plot H, which has a clear curvilinear form, has a correlation close to 1. Point out the importance of looking to see the form of the association rather than simply looking at statistical printouts.

A

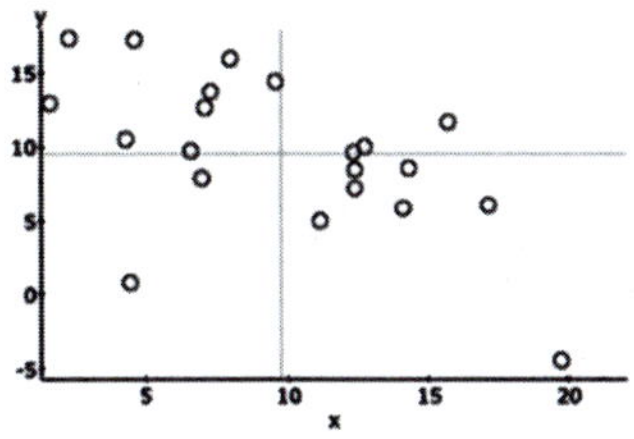

B

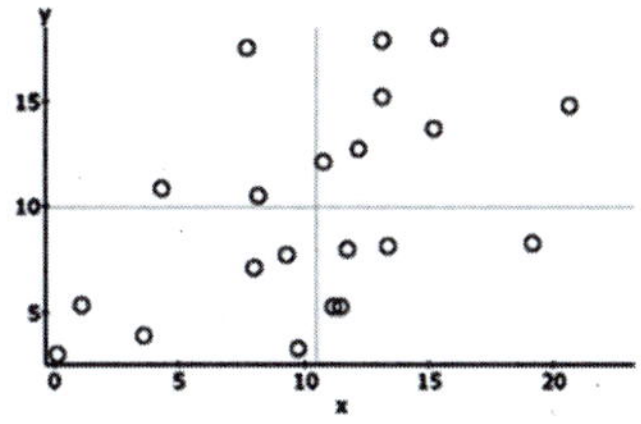

C

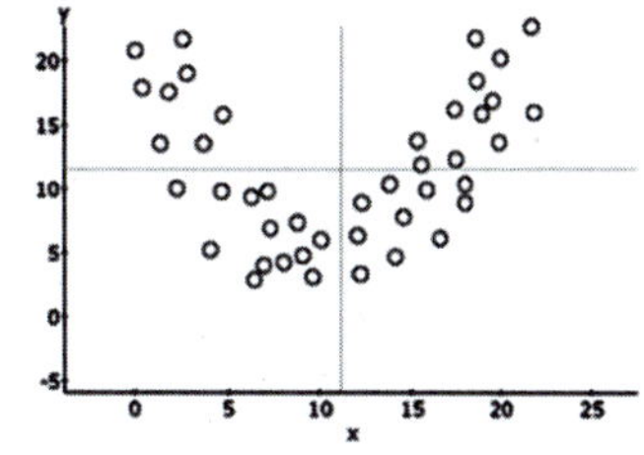

D

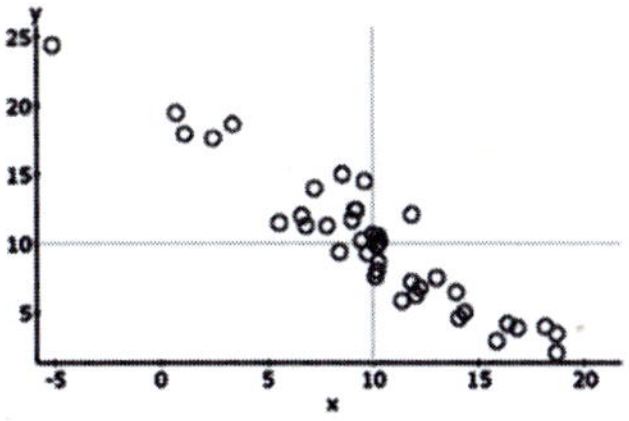

E

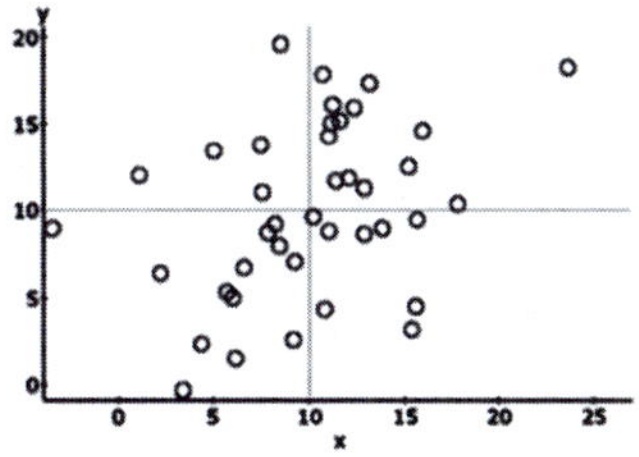

F

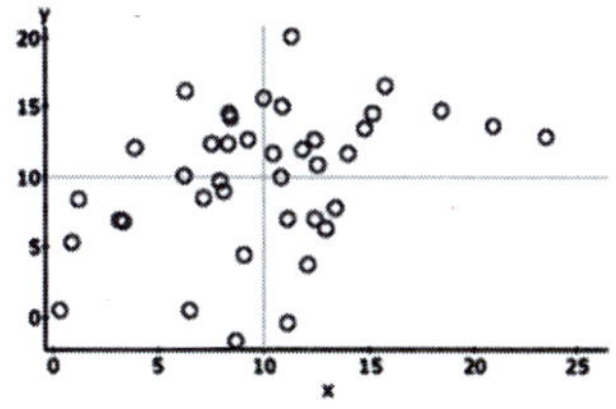

G

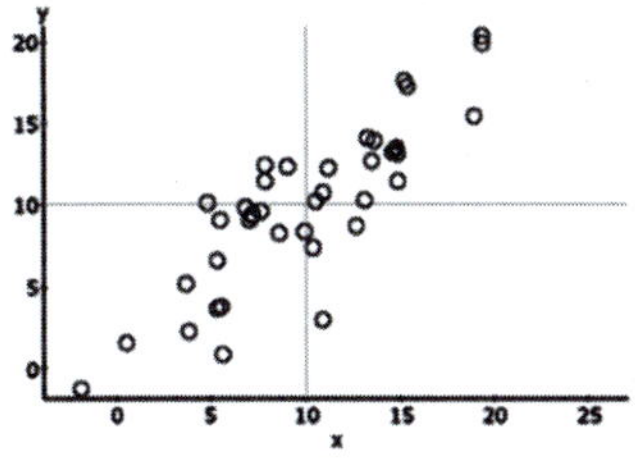

H

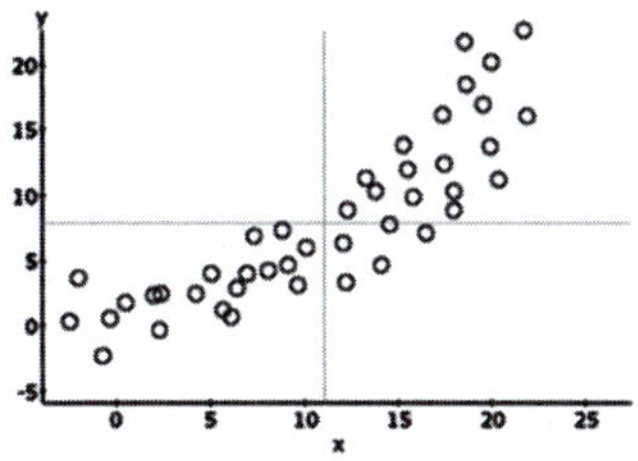

I

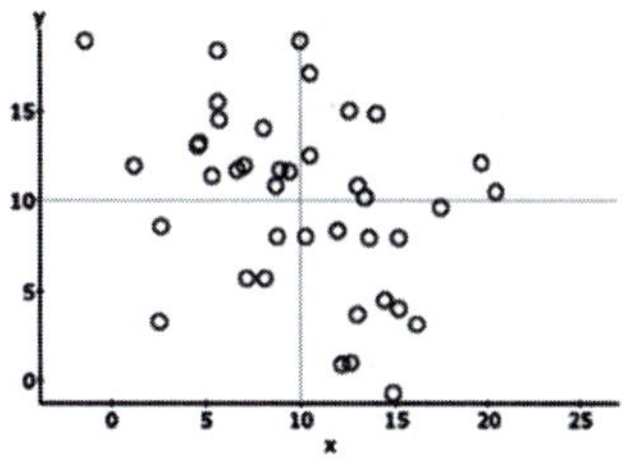

30. After plotting the residuals for a linear model, Jemila said, "Hmmm, there is a pattern in my residual plot, clearly my model is missing some aspect of the association." Draw a plot that might have led Jemila to make this statement.

Encourage students to share their ideas. The key idea is that any type of pattern in the residual plot should be cause for concern. Students are quick to find linear or curved patterns in residuals, but students should also look for "trumpeting" patterns where the size of the residuals is clearly associated with the value of the explanatory variable. (See the plots on the next page.)

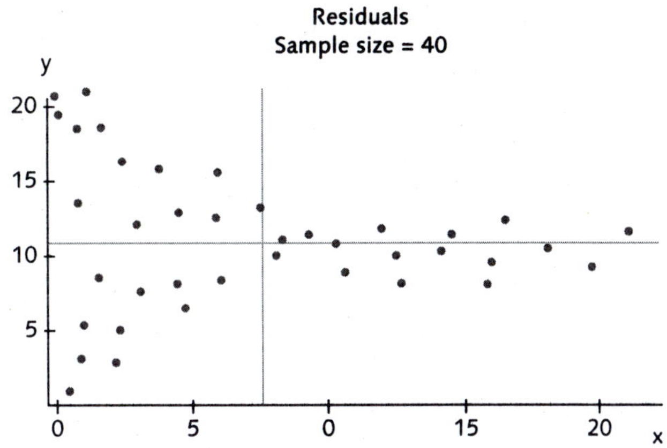
Residuals
Sample size = 40
y
20
15
10
5
0
5
0
15
20
x

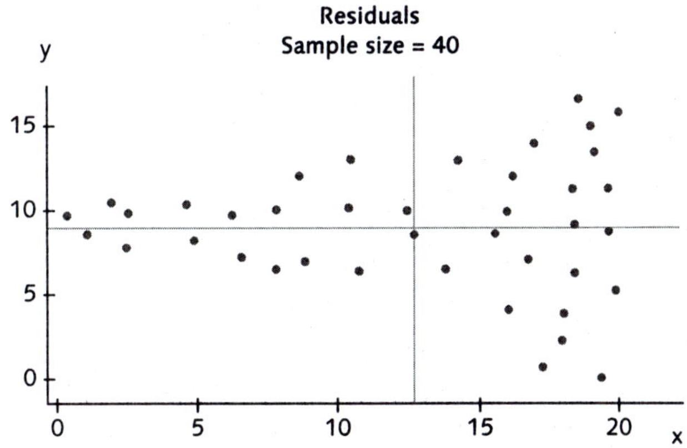
Residuals
Sample size = 40
y
15
10
5
0
0
5
10
15
20
x

Use the information that follows to answer Questions 31–34.

Z. Oologist has been gathering data on the gestation times (in days) and the longevity of 40 different animals. The following are dot plots of these two variables. The hippopotamus is a high outlier for its longevity, and the elephant is a high outlier in both variables. (Data is based on shared data sets found on www.statcrunch.com.)

(See Reproducible 65, Z. Oologist's Dot Plots.)

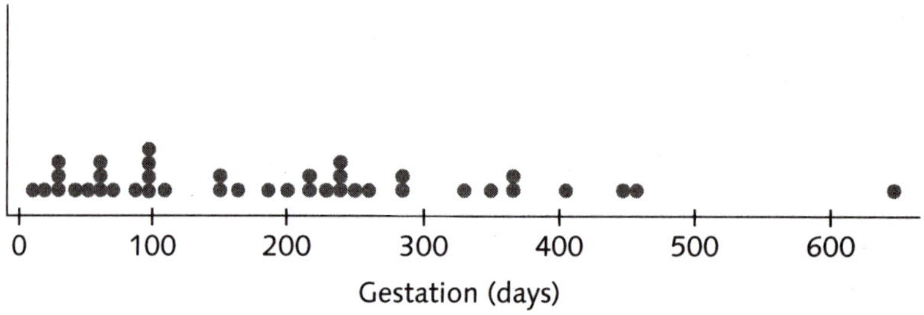

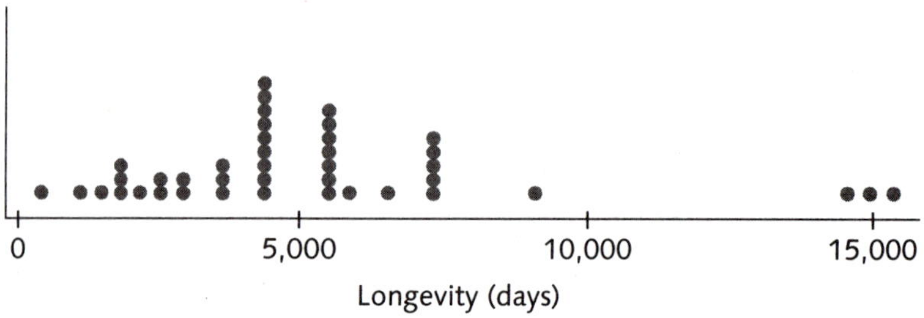

31. Read the information provided on page 240. Z. Oologist thinks that there is probably a moderate positive linear association between an animal's gestation period in days and that animal's longevity. Draw a scatterplot that would supports Z's belief.

Students have the option to choose the predictor and response variables. Note: Using predictor rather than independent variables and response rather than dependent variables helps students see the difference between statistical models and functional models.

32. Read the information provided on page 240. Z. Oologist looks at the following plot to determine whether a linear model would be appropriate for predicting longevity from gestation. What do you notice?

(See Reproducible 66, Z. Oologist's Scatterplot.)

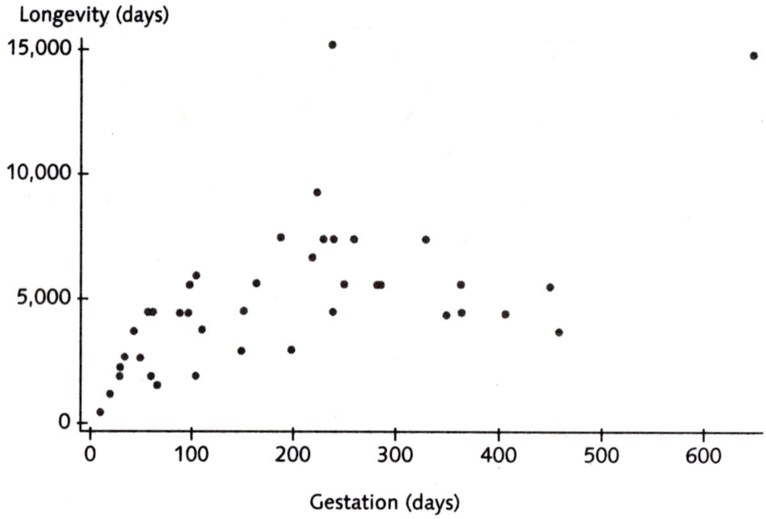

Here we want students to distinguish between outliers and influential points. The hippopotamus would be an outlier, since it would weaken the association, but not have much influence on the slope of the linear model. The elephant would have a strong influence on the correlation and the slope of the model. The table that follows shows the linear regression models and the values of R^2 and r for models that have the "points of interest" omitted.

	Model	R^2	r
With hippo and elephant	$\widehat{\text{Longevity}} = 2559.8 + 12.26 \times \text{Gestation}$	34.7%	0.589
Without hippo, but with elephant	$\widehat{\text{Longevity}} = 2415.8 + 11.74 \times \text{Gestation}$	44.7%	0.668
Without hippo and elephant	$\widehat{\text{Longevity}} = 2927.3 + 8.26 \times \text{Gestation}$	27.4%	0.523
Without elephant, but with hippo	$\widehat{\text{Longevity}} = 3021.3 + 9.26 \times \text{Gestation}$	19.8%	0.445

33. Read the information provided on page 240. Z. Oologist decided that two of the animals in the following scatterplot seem to be different. One animal weakens the strength of Z. Oologist's model, while the other animal seems to inflate the strength of the relationship. Which is which? Explain.

(See Reproducible 66, Z. Oologist's Scatterplot.)

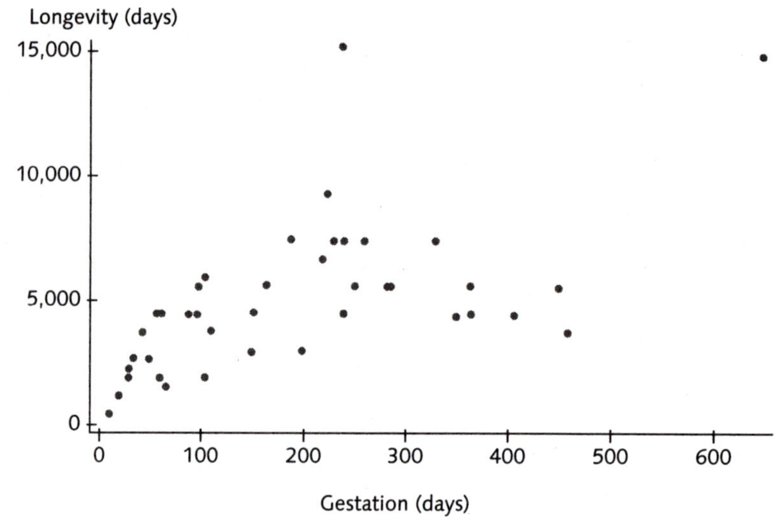

Here are the models to help students discuss their ideas. There is no fine line distinguishing an influential point from an outlier, but students should know that comparing models this way is important when analyzing data.

	Model	R^2	r
With hippo and elephant	$\widehat{\text{Longevity}} = 2559.8 + 12.26 \cdot \text{Gestation}$	34.7%	0.589
Without hippo, but with elephant	$\widehat{\text{Longevity}} = 2415.8 + 11.74 \cdot \text{Gestation}$	44.7%	0.668
Without hippo and elephant	$\widehat{\text{Longevity}} = 2927.3 + 8.26 \cdot \text{Gestation}$	27.4%	0.523
Without elephant, but with hippo	$\widehat{\text{Longevity}} = 3021.3 + 9.26 \cdot \text{Gestation}$	19.8%	0.445

Point out to students that this data is summarizing each animal with a single data point (aggregating). Ask students how the analysis might change if they had data on many animals of each type.

Students should use more descriptive variable names rather than the generic x and y.

34. See the information provided on page 240. Z. Oologist decides to omit the hippopotamus and elephant before fitting a linear model. The following are scatterplots of the data with the linear model and Residuals versus the Gestations. Based on this information, does a linear model seem appropriate? Explain.

(See Reproducible 67, Scatterplot and Linear Model.)

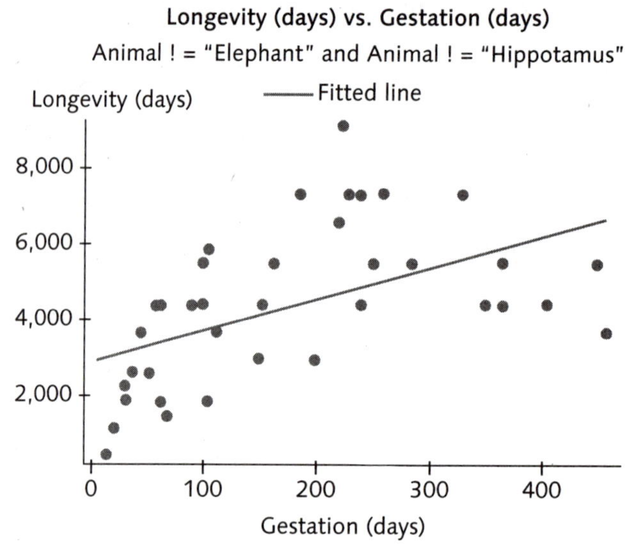

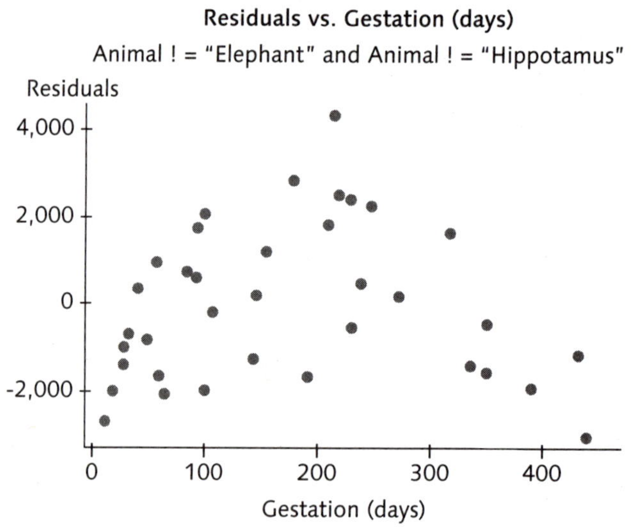

Students may notice that the residuals show an upside-down U pattern, which indicates that the errors are systematic, indicating a curvilinear relationship and so a linear model is not appropriate.

35. Define two variables that you feel have a linear association (preferably variables that can be readily measured). Then do the following:

a. Estimate the correlation coefficient.

b. Draw a sketch of what a scatterplot might look like. Be sure to label your axes.

c. Write an estimate of the linear model and interpret the slope and intercept.

Encourage students to use software (such as Desmos) to verify or amend their answers for accuracy and precision.

36. Use the given information to find a linear model for the given data.

(See Reproducible 68, Scatterplot with Data.)

Explanatory	Response	Scatterplot
2	5	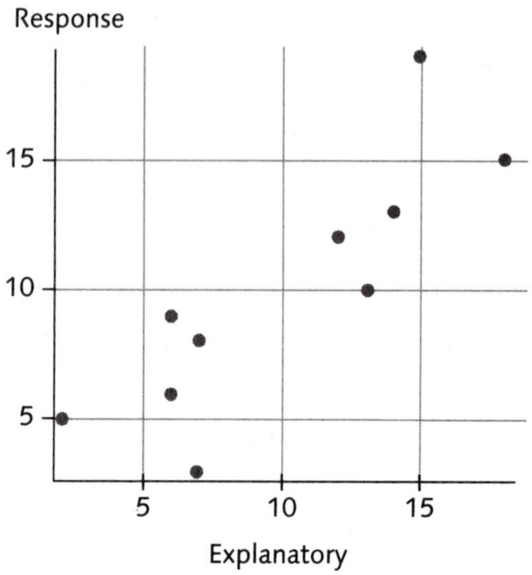
6	9	
6	6	
7	8	
7	3	
12	12	
13	10	
14	13	
15	19	
18	15	
$\bar{x} = 10$ $s_x = 5.077$	$\bar{y} = 10$ $s_y = 4.876$	Correlation = 0.844

Have students share their linear models and their choice of method. Ask students how they can assess which model is best and have them explain what criteria they are using.

Note: The least squares regression line would be $y = 1.894 + 0.811x$.

37. In statistics, we say *correlation does not imply causation.*

a. What are some examples of two variables that likely have a moderate to strong correlation, but a change in one variable does not cause a change in the other?

b. What are some examples of two variables that have both a correlation and causation relationship?

c. Is it possible for two variables to have a moderate to strong correlation and some degree of causation? Explain.

One possible answer would be that the number of doctors in a state and the number of gallons of gas consumed in a state are not causally associated. Both variables are responding to the lurking variable of population, which affects both variables. Whereas the number of miles driven does cause the amount of gas consumed in a gas-powered vehicle to increase. Here, students can discuss that there are other variables that contribute to the actual increase in fuel consumed such as type of driving (stop-and-go versus highway), the terrain, the number of passengers, and cargo. This can lead to a discussion of the advanced topic of multiple regression, which, due to software, has become much more accessible.

38. The data shown here represents 747 movies that have been released both in the United States (Domestic) and around the world (Worldwide). For each movie we have its Domestic and its Worldwide Gross earnings in millions of dollars.

(See Reproducible 69, Earnings of Domestic and Worldwide Movies.)

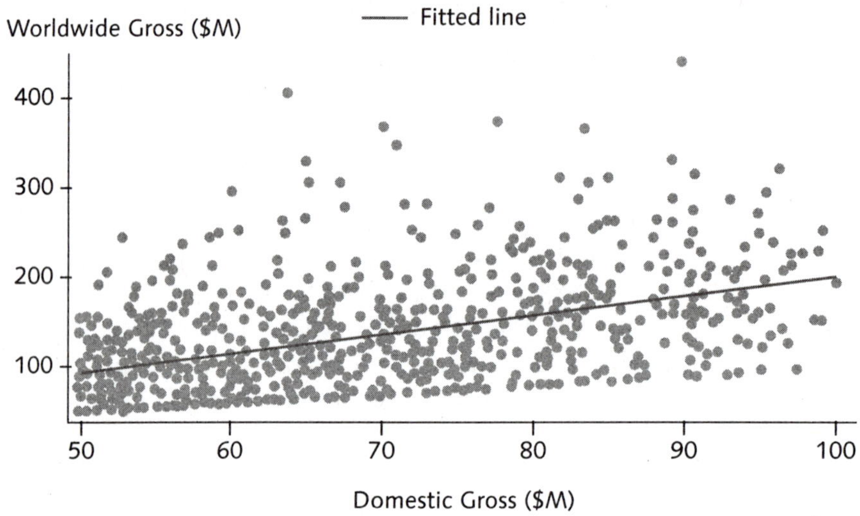

The least squares regression or "best fit" model is given by

$$\widehat{\text{Worldwide Gross}} = -16.633 + 2.179 \times \text{Domestic Gross}$$

a. Explain the purpose/use of the model.

b. State the slope and interpret its meaning in the context of the data.

c. State the intercept, interpret its meaning in context, and comment on whether it is reasonable.

d. What are some interesting features and what do you think they tell us?

e. If a movie had $70 million in domestic earnings, what would you predict would be its worldwide earnings?

f. If a movie made $200 million worldwide, can we use this model to predict its domestic earnings?

Least squares models are predictions, not functions. The LSRL model to predict Domestic Gross from Worldwide is Domestic Gross ($M) = 55.581264 + 0.099020164 Worldwide Gross ($M). So the data would suggest that worldwide earnings of $200 million would have had domestic earnings of $75.38 million.

Students who substitute 200 in the original model and then solve would predict $99.419 million. Here is one of the many areas where statistical models and traditional algebra diverge.

39. George Grapher is studying life expectancies for many countries and the number of people in that country compared to the number of television sets.

George learns that the correlation between the two variables is -0.61 (based on data from DeVeaux, Vellemen, and Bock 2018) and then concludes that a way to increase the life expectancy for a country is to send them lots of new televisions.

How did George come to this conclusion? Do you agree? Explain why or why not.

Ask students what the scatterplot would look like and discuss which variable should be the explanatory and which should be the response. This question allows students to discuss the concept of the *lurking variable*, which here would be a common response to the wealth of the people of these countries.

You can also ask students to comment on the strength of the association if the data were on individuals from these countries rather than summary statistics for the country as a whole. This is the recurring theme that aggregated data typically shows stronger trends than raw data.

Making Inferences and Justifying Conclusions

Experiences at This Level Will Help Students To

- understand and evaluate random processes underlying statistical experiments
- understand statistics as a process for making inferences about population parameters based on a random sample from that population
- decide if a specified model is consistent with results from a given data-generating process (e.g., using simulation)
- make inferences and justify conclusions from sample surveys, experiments, and observational studies

Reproducibles are available in a downloadable, printable format. See page xx for directions about how to access them.

Materials

- Results of Card-Deck-Sorting Study (Reproducible 70)

Good Questions and Teacher Notes (pages 253–263)

1. Suppose that you wanted to estimate the average word length used in William Shakespeare's written works. Three friends suggest the following:

Friend A: Grab a copy of *The Oxford Shakespeare*, an anthology of Shakespeare's works, and open to 25 random pages (using a random number generator). On each page randomly place your finger on the page and select the word closest to your finger. Find the mean and median word length.

Friend B: Open to 25 random pages (using a random number generator), and then choose a random word on that page (again, using a random number generator). Find the mean and median word length.

Friend C: Follow plan A but choose 100 random pages.

Which method should provide the most accurate estimate of Shakespeare's average word length?

Which method should be the most precise at estimating Shakespeare's average word length?

Here, students should notice that the larger words have a much greater likelihood of being selected in Methods A and C, so these two methods are biased toward a higher mean and median. Thus, Method B is the most accurate method. Since Method C has a larger sample size, it would provide the more precise estimate (lower standard error).

2. Suppose that you are told that a bag contains only red marbles and green marbles. You draw four marbles from the bag and each time the marble drawn is red.

- Abe thinks that the next marble drawn is most likely going to be red.
- Nuru thinks that the next marble drawn is most likely going to be green.
- Chris thinks that the next marble is equally likely to be red as it is to be green.

Describe what each person is assuming about the bag of marbles and the way in which they have been drawn to reach their conclusion.

One possible answer is that Abe is assuming that the marbles are replaced (i.e., with replacement) after each draw and that there are a higher number of red marbles in the bag. Nuru on the other hand is assuming that the marbles are not replaced (i.e., without replacement) and that the chance of drawing a green marble has increased since the number of red marbles has been reduced by 4. Chris may be thinking that there are an equal number of red and green marbles in the bag and they are replaced after each draw, and thus either red or green is equally likely.

Supposing that Chris is correct, ask students whether they feel that a green marble is due. This can start a good conversation about independence.

3. Suppose that Ariel and Bertha were in a coin-flipping contest.

- Ariel wins if she is the first to flip her assigned sequence HTH (Head, Tail, Head).
- Bertha wins if she is the first to flip her assigned sequence HTT (Head, Tail, Tail).

Once a player wins, the game starts over at a new first flip. For example, in the sequence of flips shown here, Bertha would win in the 8th round of flips. They play the game 20 times with Ariel winning 8 times, Bertha 11, and one draw.

Player	Flip 1	Flip 2	Flip 3	Flip 4	Flip 5	Flip 6	Flip 7	Flip 8
Ariel	T	T	T	T	H	H	H	T
Bertha	T	T	H	H	H	H	T	T

a. Based on these results, which sequence would you choose to flip for?

b. Is this evidence that Bertha's sequence (HTT) tends to happen in fewer flips than Ariel's (HTH)?

c. If they played 200 times, with Ariel winning 80 times, Bertha 110 times, and 10 draws, would this be stronger evidence that Bertha's sequence tends to happen in fewer flips?

This problem is based on a TED Talk by Peter Donnelly (https://www.ted.com/talks/peter_donnelly_shows_how_stats_fool_juries). Students should be encouraged to watch the talk, which presents three counterintuitive results.

It turns out that the sequence HTT does tend to occur in fewer flips than HTH, even though each sequence is equally probable ($\frac{1}{8}$) in any set of three coin flips. This reinforces the importance of clear definitions of random phenomena.

4. Suppose that you flipped a coin *n* times.

a. How many heads should you expect?

b. Is it always possible to get the number of heads that you expect?

c. How many heads would make you suspect that the coin is *not* fair?

d. How is your answer from part c related to the value of *n*?

Students should notice that the expected value is $\frac{n}{2}$, but if n is odd, then it is impossible to get the expected number of heads. In Questions c and d, students should notice that the number of heads depends on the size of n, but the number of heads expected increases as n increases (but at a slower rate). This discussion could also lead to focusing on the percentage of heads being more productive, since the expected range of percentages "around 50 percent" should decrease as the value of n increases. This supports students' intuition that estimates become more precise as the sample size n increases.

5. Of the last 100 days that the weather person on the local news predicted there was a 70 percent chance of rain, it rained on only 60 of those days. Should the local news be looking for a new weather person? What if it rained on 80 of those days?

Here, the chance that 60 or fewer days of rain would be observed is 2 percent or about 1 in 50, so there is some, but not overwhelming, evidence that the weather person has underperformed. If it had rained on 80 of those days, that would be slightly more evidence since the likelihood of that happening by chance is only 1.6 percent. Depending on whether students have learned binomial probability will drive the depth of the conversation.

Ask students if their answers change if instead it was only 10 days (6 days of rain) or 1,000 days and 600 rainy days.

6. When comparing the effects of two treatments, which of the following aspects of a study are most important? Explain your choice.

- the size of the difference
- the variability within each treatment group
- the sizes of the samples
- the design of the study

Collect data on which aspect your students think is most important. The key here is that all four are crucial in forming a useful comparison. Have students discuss scenarios that undermine each option (e.g., in looking at a large difference between the two treatments, students can point to biased sampling, small sample sizes, or large variability as ways that would render the large difference to be inconclusive).

7. Imagine you were interested in investigating whether people, on average, can sort a standard deck of cards into two piles—with hearts and diamonds in one pile and clubs and spades in the other—faster than they can sort them into two piles with hearts and spades in one pile and clubs and diamonds in the other:

a. How might you go about designing an experiment to investigate this question?

b. Suppose a well-designed study revealed the following results.

(See Reproducible 70, Results of Card-Deck-Sorting Study.)

Sample	Mean	Standard Deviation
Sort Time (in seconds) Hearts and Diamonds versus Clubs and Spades	53.836667	3.3031853
Sort Time (in seconds) Hearts and Spades versus Clubs and Diamonds	62.123333	5.7704881

Would these results be considered strong evidence that people are, on average, faster at sorting a standard deck into hearts and diamonds in one pile, clubs and spades in the other? Explain.

c. What information would help answer the question with more confidence?

There are a few areas of potential focus here. The first is that the number of participants is crucial in determining the strength of the evidence on whether there is a significant difference. For example, if there were 5 subjects in each group, the *P* value in a two-sample *t*-test is 0.015, so as long as the experiment had a valid design, this would be evidence that the hearts and diamonds sorting is faster for people.

Students may question the design, saying that the people selected for the hearts and diamonds group are just faster sorters in general, which confounds the results. They may also suggest that a matched pairs design may be better, so that each person acts as their own control. In the matched pairs, they would need to randomly assign each subject which type of sorting they did first, and then the variable of interest is each subject's difference in sorting times between hearts and diamonds versus hearts and spades. You could then ask what results would support the claim and what results would fail to support the claim.

8. In an attempt to study the relationship between job callbacks and email address names, eStat Company plans to send online resumes to a large number of online job postings using two different resume types—high quality and low quality—and two different types of email addresses—ones containing a fictitious actual name such as john.smith@ . . . , and ones containing a gimmick name, such as bedazzled077@ . . . . How might you design such a study to obtain reliable statistical data?

There are two factors involved in this study, each with two levels: the quality (high and low) and the email type (name and gimmick), which means there are four treatment groups. One possible study design would be to randomly assign each online job posting one of the four treatments. Then record whether the job posting gave a callback. You could then conduct an analysis of the percentage of callbacks for each treatment group.

You could ask students to discuss the number of job postings needed to observe a difference. Depending on the background of the class, you could ask students what type of analysis they might conduct. For example, they could look at bar graphs and segmented bar graphs. Also they could conduct two-proportion z-tests for each factor, or chi-square tests for independence among the four treatment groups.

9. **Based on an article in *Chance* magazine (http://chance.amstat.org), it has been suggested that while flipping a coin is a fair method of deciding between two options, spinning and tipping may not be.**

   a. Suppose that Robin obtained only 45 percent heads when spinning a penny. How many spins would be required for this to be considered proof that spinning a penny is not as fair as flipping?

   b. Suppose that Robin obtained only 69 percent heads when tipping a penny. How many tips would be required for this to be considered proof that tipping a penny is not as fair as flipping?

Spinning involves spinning the coin, on edge, on a smooth surface until it lands either heads up or down. Tipping involves standing a penny on edge on a smooth surface and gently tapping the surface until the penny tips over. Students should probably find that 100 spins or tips would not be much proof but that 10,000 certainly would be proof. This question asks them to find a place in between where they would shift from "not much proof" to "significant proof" (e.g., students could be convinced that only 40 tips would suffice to conclude that 69 percent heads implies tipping is *not* a fair method for deciding, while they would need about 400 spins before concluding that 45 percent heads implies that spinning is unfair).

10. Suppose that eStat Company (from Question 8) implemented a well-designed study and discovered that resumes with an actual name in the email address received 50 percent more callbacks than those using a gimmick name in their email addresses.

a. Create a two-way table that would support the given finding and discuss whether the result would be considered statistically significant.

b. Create another two-way table that would support the given finding and discuss whether this new result would be considered stronger (or weaker) evidence than your first set of data.

One possible solution would be:

In the results that follow here, there were 50 percent more callbacks, yet the callback percentages are similar (60 versus 57), so while the results support the headline, the differences in the number of callbacks is *not* significant.

Email address	Number of Callbacks	Number of Resumes
Actual name	30	50
Gimmick	20	35

Now look at these results:

Email address	Number of Callbacks	Number of Resumes
Actual name	30	50
Gimmick	20	50

Here, the percentage of callbacks is 60 versus 40, which would be significant (the *P* value of the two-proportion test is 0.023).

Encourage students to explore the effects of changing the callback rates using different sample sizes to get a better sense of how the *50 percent more* statement could be true but possibly not significant.

Another possible solution is shown here; little or no evidence that the actual name made a resume more likely to be called back, using 500 resumes with a 7 percent callback rating for the gimmick name and 10.5 percent for the actual name.

Status/Name	Actual Name	Gimmick Name	Total
Callbacks	27	17	44
No Callbacks	233	223	456
Total	260	240	500

Here, there is some evidence that the actual name made a resume more likely to be called back, using 1,000 resumes with a 7 percent callback rating for the gimmick name and 10.5 percent for the actual name.

Status/Name	Actual Name	Gimmick Name	Total
Callbacks	55	34	89
No Callbacks	465	446	911
Total	520	480	1,000

This table shows strong evidence that the actual name made a resume more likely to be called back, using 1,000 resumes with a 14 percent callback rating for the gimmick name and 21 percent for the actual name.

Status/Name	Actual Name	Gimmick Name	Total
Callbacks	109	67	176
No Callbacks	411	413	824
Total	520	480	1,000

Conditional Probability and the Rules of Probability

Experiences at This Level Will Help Students To

- understand independence and conditional probability and use them to interpret data
- recognize and explain the concepts of conditional probability and independence in everyday language and everyday situations
- describe events as subsets of a sample space (the set of all possible outcomes) using characteristics (or categories) of the outcomes, or as unions, intersections, or complements of other events ("or," "and," "not")
- construct and interpret two-way frequency tables of data to determine if events are independent and to approximate conditional probabilities
- use the rules of probability to compute probabilities of compound events

Reproducibles are available in a downloadable, printable format. See page xx for directions about how to access them.

Materials

- Table of Qualities Important to Fourth- through Sixth-Grade Students (Reproducible 71)
- Probability Venn Diagram (Reproducible 72)

Good Questions and Teacher Notes (pages 265–274)

1. Consider a standard deck of cards. Define two events, *A* and *B*, and then list the events that correspond to

- *A* or *B*
- *A* and *B*
- Not *A*
- Not *B*
- Not (*A* or *B*)
- Not (*A* and *B*)

Have students post their answers for each subset and look for similarities.

A possible answer is let event *A* = the card is an ace, and event *B* = the card is a face card; then the set of events for each is shown here:

- *A* or *B*: {the four aces and the twelve face cards}
- *A* and *B*: {}
- Not *A*: {the 48 non-aces}
- Not *B*: {the 40 non-face cards}
- Not (*A* or *B*): {a 2, 3, 4, . . . 10 card, any suit}
- Not (*A* and *B*): {any card}

Ask what it means if "*A* and *B*" is an empty set. Discuss mutually exclusive/disjoint events.

Ask what it means if "*A* or *B*" is the same as the sample space for drawing cards. Here, the two events must be complementary.

Consider the following for Questions 2–4: Roll two fair dice, one six-sided {1, 2, 3, 4, 5, 6} and the other four-sided {1, 2, 3, 4}, and find the product of the two results.

2. Write a sample space in which the outcomes are equally likely. Write a sample space in which the outcomes are *not* equally likely.

Have students share how they chose to organize the information (e.g., trees, tables, lists). Discuss how each display can be used to examine "likelihood."

3. Suppose that

- Angie is looking for outcomes in which the result on the four-sided die is at least the result on the six-sided die *and* the product is a perfect square.
- Bela is looking for outcomes in which the result on the four-sided die is at least the result on the six-sided die *or* the product is a perfect square.

Whose subset of outcomes is larger? Explain.

Careful reading is a key component to probability. Here Angie has only 5 possible outcomes (4-sided, 6-sided) {(1, 1), (4, 1), (2, 2), (3, 3), (4, 4)} whereas Bela has 10 outcomes, Angie's outcomes plus {(2, 1), (3, 1), (3, 2), (4, 2), (4, 3)}.

4. Define two events such that $P(A \text{ or } B) > P(A \mid C)$.

One possible solution is A = the 4-sided die is even, B = the 6-sided die is even, and C = the sum of the roll is even.

$$P(A \text{ or } B) = \frac{18}{24} = 0.75 > P(A \mid C) = \frac{6}{12} = 0.5$$

5. AB's teacher asked her the following question:

A bag of M&M candies contains 50 candies: 5 blue, 11 brown, 7 green, 14 orange, 3 red, and 10 yellow. Two M&Ms are drawn from the bag without replacement. If ***A*** is the event that a blue M&M is drawn and ***B*** is the event that an orange M&M is drawn, what is the probability that ***A*** and ***B*** occur together?

AB said, "The probability is 1/50 because there are 50 candies and only 1 way for a blue and orange to be drawn together." Do you agree or disagree with AB's answer? Explain.

No, there are two ways to have events A and B occur together. Draw a blue and then an orange *or* draw an orange and then a blue. So

$$P(A \text{ and } B) = P((A1 \text{ and } B2) \text{ or } (A2 \text{ and } B1))$$

$$P(A \text{ and } B) = P(A1 \text{ and } B2) + (A2 \text{ and } B1)$$

$$P(A \text{ and } B) = P(B2 \mid A1) + P(A2 \mid B1)$$

$$= \frac{5}{50} \times \frac{14}{49} + \frac{14}{50} \times \frac{5}{49} = \frac{2}{35}$$

6. JR's teacher asked him the following question:

A bag of M&M candies contains 50 candies: 5 blue, 11 brown, 7 green, 14 orange, 3 red, and 10 yellow. You are going to draw two at random. Suppose we define ***A*** as the event that 1 red and 1 orange M&M are drawn and ***B*** as the event that 1 yellow and 1 green M&M are drawn. Is the probability that ***A*** occurs and the probability that ***B*** occurs equal? Explain.

JR wrote, "Yes, these two probabilities are equal since there are 17 M&Ms that are red or orange and there are 17 M&Ms that are yellow or green." Do you agree or disagree with JR's answer?

No; here the common incorrect response given is about the equal number of candies in each group, 17. However, the two probabilities are different as shown:

$P(A) = P(R1 \text{ and } O2) + P(O1 \text{ and } R2)$

$P(A) = P(R) \cdot P(O|R) + P(O)$

$P(A) = \frac{3}{50} \cdot \frac{14}{49} + \frac{14}{50} \cdot \frac{3}{49}$

$P(A) = \frac{6}{175}$

$P(B) = P(Y1 \text{ and } G2) + P(G1 \text{ and } Y2)$

$P(B) = P(Y) \cdot P(G|Y) + P(G)$

$P(B) = \frac{10}{50} \cdot \frac{7}{49} + \frac{7}{50} \cdot \frac{10}{49}$

$P(B) = \frac{2}{35}$

7. A randomly selected set of students in grades 4–6 from three large school districts were asked whether good grades, popularity, or athletic ability was the most important goal to them. Here is a two-way table summary of grade level versus goal.

(See Reproducible 71, Table of Qualities Important to Fourth- through Sixth-Grade Students.)

		Grade			
		4	5	6	*Total*
Goal	Grades	63	88	96	247
	Popular	31	55	55	141
	Sports	25	33	32	90
	Total	119	176	183	478

If we were to randomly select one student from this study for a follow-up interview, what questions could you ask that would illustrate each of the probabilities listed?

- $P(A)$
- $P(B)$
- $P(A \text{ or } B)$
- $P(A \text{ and } B)$
- $P(A \mid B)$
- $P(B \mid A)$

When discussing this question, be sure that students clearly define two separate events, *A* and *B*, and explain whether they are independent and/or mutually exclusive—two important concepts for which students often confuse the terms.

8. A bag contains n red balls numbered 1 to n and m green balls numbered 1 to m. If you randomly select two balls from the bag:

a. Describe how this can be done such that the results of each ball drawn are independent and write an expression for the probability that both balls are green.

b. Describe how this can be done such that the results of each ball drawn are *not* independent and write an expression for the probability that both balls are green.

c. Suppose the balls are drawn without replacement. What must be true about n and/or m in order for the results of each ball drawn to be assumed to be essentially independent?

This question gets at the 10 percent condition in hypothesis tests, which assumes that the sample size cannot exceed 10 percent of the population in order to assume independence, which is an essential assumption in computing the spread of a sampling distribution.

9. **Define two events, *A* and *B*. Fill in the elements of the following Venn Diagram and then find the probabilities listed.**

(See Reproducible 72, Probability Venn Diagram.)

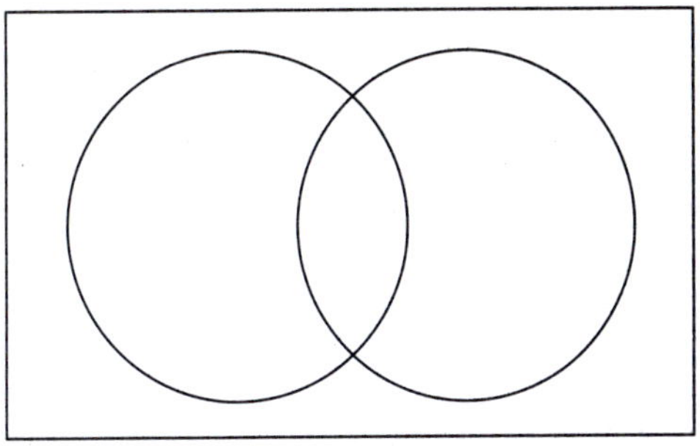

a. $P(A)$

b. $P(B)$

c. $P(A \text{ or } B)$

d. $P(A \text{ and } B)$

e. $P(A \mid B)$

f. $P(B \mid A)$

Students need to clearly define two separate events here. Ask students if *A* and *B* are independent. Ask students if *A* and *B* are disjoint or mutually exclusive. Have students state the contextual meaning of each probability.

10. Brent was interested in the pet preferences of the students at his school. First, he found the probability that a randomly selected student had at least one dog as a pet. Then he found the probability that a randomly selected student had at least one cat as a pet. When his teacher heard about his study, she asked Brent for the probability that a randomly selected student had at least one dog or cat. Brent quickly added these two probabilities together and seemed confused when his answer was larger than 1. Explain what happened.

If necessary, you can scaffold this question by asking students to provide possible values of $P(A)$, $P(B)$, $P(A \text{ or } B)$, and $P(A \text{ and } B)$ and explain what they mean in this situation.

Note: The Venn Diagram introduced in the previous question (Reproducible 72) might be very useful here.

11. Consider a standard deck of cards.

a. Consider the random phenomenon of drawing a single card. Define two events, *A* and B, and then write out the relationship: $P(A \text{ or } B) = P(A) + P(B) - P(A \text{ and } B)$ and interpret the statement.

b. Consider the random phenomenon of drawing two cards. Define two events, *A* and *B*, and then write out the relationship: $P(A \text{ or } B) = P(A) + P(B) - P(A \text{ and } B)$ and interpret the statement.

Students often struggle to define events clearly. In part a, they should define events like *A* = the card is a heart; *B* = the card is an ace. In Part b, they have more options such as *A* = the first card is a heart; *B* = the second card is a heart.

Ask students whether their answers for part b would change if the cards were drawn with replacement versus without replacement.

12. Consider a standard deck of cards.

a. Consider the random phenomenon of drawing a single card. Define two events, A and B. Can we use the relationship $P(A)P(B \mid A) = P(B)P(A \mid B)$ to find $P(A \text{ and } B)$? Explain.

b. Consider the random phenomenon of drawing two cards. Define two events, A and B. Can we use the relationship $P(A)P(B \mid A) = P(B)P(A \mid B)$ to find $P(A \text{ and } B)$? Explain.

Yes to both. A possible answer to the first problem would be: if A = the card is a face card, and B = the card is a heart, then

$$P(A \text{ and } B) = P(A)P(B \mid A) = P(B)P(A \mid B)$$

$$P(A \text{ and } B) = \frac{12}{52} \times \frac{3}{12} = \frac{13}{52} \times \frac{3}{13} = \frac{3}{52}$$

For the second problem, a possible answer is A = first card is a face card, and B = second card is a face card.

$$P(A \text{ and } B) = P(A)P(B \mid A)$$

$$P(A \text{ and } B) = \frac{12}{52} \times \frac{11}{51} = \frac{11}{221}$$

The two problems address two separate concepts. Since there is only one card drawn in the first, the question is whether the events are disjoint (i.e., $P(A \text{ and } B) = 0$) or not. In the second, the question is whether the events are independent (i.e., $P(A \text{ and } B) = P(A)P(B)$, which means that $P(B \mid A) = P(B)$ and $P(A \mid B) = P(A)$).

13. Can you create two events that are both disjoint and independent?

How about two events that are neither disjoint nor independent?

You may want to scaffold this question by focusing on a particular random phenomenon. For example, if drawing two cards from a standard deck of 52 cards, the answer to the first question is no, since disjoint events mutually exclude each other, which means they are not independent. The answer to the second question is yes. A possible solution is A = the cards are both red, B = the cards are both hearts. They are not disjoint since $P(A \text{ and } B) = \frac{13}{52} \times \frac{12}{51} = \frac{1}{17}$ and they are not independent since $P(A \mid B) = 1$ which is not equal to $P(A) = \frac{25}{102}$.

14. Blaise fills a bag with 15 marbles that are either red or green. Bernie draws two marbles from the bag without replacement. Blaise was not paying close attention and only saw that the second marble drawn was green. He knows, based on the marbles he put in the bag, that P(first marble is green | second marble is green) $= \frac{2}{7}$.

Can you determine how many of each marble Blaise placed in the bag originally?

The answer is yes, Blaise put 5 green and 10 reg marbles in the bag. A good way for students to organize/visualize the information is with a tree diagram.

Formally, this problem draws on Bayes' theorem, which applies the multiplication rule ($P(A \text{ and } B) = P(A)P(B \mid A) = P(B)P(A \mid B)$).

$$P(G_1|G_2) = \frac{P(G_1 \text{ and } G_2)}{P(G_2)} = \frac{P(G_1 \text{ and } G_2)}{P(G_1 \text{ and } G_2) + P(G'_1 \text{ and } G_2)}$$

Using Probability to Make Decisions

Experiences at This Level Will Help Students To

- calculate expected values and use them to solve problems
- define a random variable for a quantity of interest by assigning a numerical value to each event in a sample space; graph the corresponding probability distribution using the same graphical displays as for data distributions
- calculate the expected value of a random variable; interpret it as the mean of the probability distribution
- develop a probability distribution for a random variable defined for a sample space in which theoretical probabilities can be calculated; find the expected value

Good Questions and Teacher Notes (pages 276–284)

1. Kapono and Farhad are playing a game that involves flipping a penny and a nickel. When Kapono flips the coins, if they are both heads, Farhad has to pay Kapono $20; if they are both tails, Kapono has to pay Farhad $10; if the coins show one head and one tail, then Kapono pays Farhad $5. Define a random variable for Kapono's earnings per game and draw the corresponding probability distribution. Is this a fair game? Why or why not?

The game is fair in that the expected payout to Kapono is $0 per game even though each game results in Kapono either winning $20 or losing either $10 or $5. Ask students to explain what the expected value means for a single game. What does it mean if Farhad and Kapono played hundreds of rounds?

As an extension, ask students to devise another payout combination that would result in a fair and a not fair game. For example, if the payout for two heads was $30, then Kapono would have an expected value of winning $3 per game even though he loses money in 75 percent of the games played.

As another extension, survey the class to see whether they would rather be Kapono or Farhad . . . if any students are familiar with behavioral economics, this can be a fruitful discussion.

2. Random variables X and Y each have the same set of possible values. Can you create a probability distribution for each so that $E(X) > 2E(Y)$?

A possible solution is shown here:

X	$P(X)$
1	0.1
2	0.05
3	0.05
4	0.15
5	0.65
$E(X)$	4.2

Y	$P(Y)$
1	0.6
2	0.1
3	0.1
4	0.1
5	0.1
$E(Y)$	2

3. On a multiple-choice test,

- Students earn 1 point if they answer correctly.
- Students lose $\frac{1}{4}$ point if they answer incorrectly.
- Students earn 0 points for questions left blank.

Describe a random variable and its probability distribution for the points earned on a question when a student randomly guesses. Under what conditions is guessing expected to help? Hurt? Have no expected effect?

Here, if there are 5 answer choices, this scoring has an expected value of 0. If there are only 4 answer choices, the expected value is +0.0625. With 3 answer choices the expected value is 0.17.

Ask students if an expected value of 0 means that the likelihood of improving your score through guessing is the same as the likelihood of harming your score. It is not, since you are four times as likely to harm your score. This can create a good discussion of risk versus reward.

4. Stephanie and Wilma are designing a new computer video game. They decide that they want players to be able to complete a particular task in 5 mouse clicks or fewer. If they design it so that only 10 percent of players can complete the task in 1 click, create a random variable and probability distribution for the number of clicks a (random) player will require to complete this task, if they want the average player to require 4 or more clicks.

Why might Stephanie and Wilma want the average player to click 4 or more times?

One possible solution is shown where 4.15 clicks are expected.

Clicks	1	2	3	4	5
P(Clicks)	0.1	0.05	0.1	0.1	0.65

Ask students to share their solutions. Then discuss how their choices might affect the player's experience.

As an extension, ask students to compute the expected value.

5. On a multiple-choice test with 5 answer choices,

- Students earn 1 point if they answer correctly.
- Students lose $\frac{1}{4}$ point if they answer incorrectly.
- Students earn 0 points for questions left blank.

Under what conditions might it be better for a student to randomly guess instead of leaving a question blank?

You can ask students whether the expected value is more important or whether the probability of improving their score is higher than the chance that they hurt their score. For example, if we suppose there are 5 such questions, and students cannot eliminate any answer choices, they have an expected value of 0; that is, the guessing should neither hurt nor help, but the students have a 33 percent chance of lowering their score and only a 27 percent chance of improving it. Whereas, if they could eliminate one choice per question, then the expected value is 0.31 and the chance of lowering their score is 24 percent and the chance of raising their score is 37 percent.

This can be extended to guessing on 10 questions and discussing the random variable representing the points earned on 10 guessed questions and its associated probability distribution.

6. A circle with a spinner is divided up into an odd number, n, of congruent sectors (through the center) numbered 0 thru $n - 1$.

- The sector labeled 0 is colored green.
- The remaining even-numbered sectors are red.
- The odd-numbered sectors are black.
- If the spinner lands on a red sector, you win \$1, otherwise, you lose \$1.

Describe the expected value of this game as the value of n increases. Will the game ever be considered fair?

Here students should notice that the game will never be fair since the green 0 sector tips the scales against them, but that it gets closer to fair as the number of sectors increases. For example, with 3 sectors they expect to lose 33 cents per game; for 11 sectors they expect to lose 9 cents per game, and for 37 sectors (like some roulette wheels) they expect to lose 2.7 cents per game.

You can also ask students whether they would rather choose to play the described game 100 times or play once for a \$100 payout if the spinner lands on red, but lose \$100 otherwise. What are the similarities and differences between the two choices?

7. A circle with a spinner is divided up into an odd number, n, of congruent sectors (through the center) numbered 0 thru $n - 1$.

- The sector labeled 0 is colored green.
- The remaining even-numbered sectors are red.
- The odd-numbered sectors are black.

There are a few options for this game of chance.

Option 1: If the spinner lands on a red sector, you win \$1, otherwise, you lose \$1.

Option 2: Select a numbered sector, p. If the spinner lands on a that sector, you win $\$(n - 2)$, otherwise, you lose \$1.

Which is the better option?

Does the value of n matter?

Here, both options have the same expected value of $\$(-\frac{1}{n})$ per game, but Option 2 has a larger variability, which means that you could win more in a single game.

8. A circle with a spinner is divided up into an odd number, n, of congruent sectors (through the center) numbered 0 thru $n - 1$.

- The sector labeled 0 is colored green.
- The remaining even-numbered sectors are red.
- The odd-numbered sectors are black.

You are trying to assign payouts that will entice people to play but will give you a positive expected value (and the players a negative expected value).

How many sectors should the circle have?

What will you pay out for what outcomes?

What is your expected value?

What does the expected payout mean in the short run? The long run?

One possible solution is to have 11 sectors numbered 0 through 10. Players pay \$1 to spin, select a number, p {0, 1, 2, . . . , 10}, and if the spinner comes on their selected number, then they win \$$p$. If players only chose 10, then the expected value is 0. If players select any other combination of numbers, then the expected value is positive. In the short run, players will have actual payouts either significantly higher or slightly lower than expected, but over the long run, their per game payout will approach the expected payout.

9. If we define a random variable to be the product of the results on each die, how do you think that the mean and median values of this random variable compare? Why? How can those change if the dice are not fair?

For Questions 9 and 10, consider rolling two fair dice; one is six-sided {1, 2, 3, 4, 5, 6} and the other is four-sided {1, 2, 3, 4}.

Here, the mean product is 8.75 and the median product is 6. Students should recognize that the values of the random variable are skewed to the right. If students struggle to answer this question, encourage them to find the probability distribution. Then they can try different distributions to explore the effects of "loaded" dice.

10. What is the expected median value of the random variable? What is the expected mode value of the random variable? What do you notice?

Students should notice the right skew to the distribution to be able to conclude that the mean should be higher than the median.

11. Suppose you roll two fair six-sided dice. The points you get are equal to the absolute value of the difference of the two dice. Create a probability distribution. What is the expected number of points? What is the most likely result?

This type of question helps students to shift from probability ideas to random variables.

You could also ask, *What if one die were red and the other green, and the score were the green minus the red?*

Ask students to generalize for *n*-sided dice and describe the probability distribution as *n* increases.

12. **Suppose that your history teacher decides to give a five-question multiple-choice (each with five options) pretest on a topic that you know nothing about. She plans to employ an interesting scoring method.**

- You get 20 points for each correct response.
- You get 5 points for each question you leave blank (indicating you don't know the answer).
- You lose 2 points for each incorrect response.

You want to perform your best. Should you guess on each question or employ a different strategy?

If your teacher made one of the responses on each question clearly wrong, would you change your strategy?

There are a variety of possible reasonings here since the expected value is one approach, but students can also look at the probability distribution of various outcomes to inform their strategy. For example, the expected value via guessing here is 12 points, but you have just over a 1 in 4 chance of scoring 34 or better rather than settling for the 25 points you would earn by leaving each question blank.

If you can eliminate one option, the expected value is 17.5 points and you have better than a 1 in 3 (36.7 percent) chance of earning 34 or more points. If you could eliminate 2 options, then guessing gives you better than 50-50 of exceeding the 25 points you'd get by leaving all 5 blank.

Have students try to devise a scoring point assignment that makes the expectation the same whether you guess or leave them blank.

CHAPTER 9

GOOD QUESTIONS FOR PRE-CALCULUS AND BEGINNING CALCULUS

Calculus requires students to combine what they know about symbolic algebra, functions, and geometry to explore change (such as rates of change) and accumulation (such as integration). For example, students extend the idea of slope from linear algebra to a dynamic view of slope of a nonlinear function at a point, as well as a functional representation of the instantaneous change over x referred to as the derivative function. In addition, students use their geometric tools to find areas of complex regions and volumes of revolution. Students in pre-calculus and calculus also grapple with challenging ideas around infinity, such as how infinitely small quantities can add up to something other than 0, and limits, such as how a function can approach a value L as x approaches a even if the function is not defined when $x = a$. The fundamental theorem of calculus pulls all of these concepts together, establishing a complex relationship between area under a curve and the derivative of a function.

Limits and Continuity

Experiences at This Level Will Help Students To

- recognize that evaluating a limit using a table is problematic, and that a graph is more reliable
- understand that limits exist when they are bounded. Therefore, limits to infinity do not exist
- define a limit
- define continuity

Reproducibles are available in a downloadable, printable format. See page xx for directions about how to access them.

Materials

- Kayla's Table (Reproducible 73)
- Limit Statements from Graph (Reproducible 74)

Good Questions and Teacher Notes (pages 287–292)

1. Kayla calculated the limit of $f(x) = \sin\left(\frac{\pi}{x}\right)$ as x approaches 0 from the right by evaluating $f(x)$ as shown in the table. However, she does not understand how this connects to the graph of $f(x)$. What happened? Explain.

(See Reproducible 73, Kayla's Table.)

x	*f(x)*
0.1	0
0.01	0
0.001	0
0.0001	0
0.00001	0

This problem shows the flaw with using tables to estimate limits. The selected x-values do suggest the limit is 0, but the way the period of the function depends on x makes this misleading. When students graph the function (in radians or degrees), they will see instead how the graph increasingly oscillates between -1 and 1 as x approaches 0, which should help students to conclude that the limit does not actually exist. Ask students why the table is misleading, and how they could select x-values that would not be misleading. Students should recognize that any x-values that approach 0 but that do not result in a multiple of 2π will work.

2. Andrew and Kevin are having an argument about the limit of $y = \frac{1}{x}$ as x approaches 0 from the right.

Andrew claims that the limit does not exist.

Kevin says that since y continues to increase without bound as x approaches 0 from the right, the limit is positive infinity. What do you think? Explain your thinking.

They are both right. This problem brings up the opportunity to have students think about what it means for a limit to exist. Although the graph allows a student to see, predict, and describe what happens to the value of y as x increases, the limit is unbounded and therefore does not exist. That is, for a limit to exist, it must be bounded, and therefore finite.

3. When $g(x) = \dfrac{2x^2 - x - 6}{5 - x}$, what are the limits here?

a. $\lim\limits_{x \to -\infty} g(x)$ b. $\lim\limits_{x \to -\frac{3}{2}} g(x)$ c. $\lim\limits_{x \to 5} g(x)$

d. $\lim\limits_{x \to +\infty} g(x)$ e. $\lim\limits_{x \to 0} g(x)$ f. $\lim\limits_{x \to 5^+} g(x)$

This problem requires students to look at various parts of a function and interpret its limit. They include limits that examine end behavior, limits at a point of the function, and limits involving a vertical asymptote. Students can see why, in the case of $\lim\limits_{x \to 5} g(x)$, the limit does not exist because the right- and left-handed limits do not agree. On the other hand, $\lim\limits_{x \to 5^+} g(x) = -\infty$, which also does not exist because infinity is unbounded.

4. What function could be described with the limits given here? Can you find more than one?

$\lim_{x \to 0} g(x)$ does not exist $\qquad \lim_{x \to 2} g(x) = \frac{1}{2} \qquad \lim_{x \to -\infty} g(x) = 0$

This puzzle requires students to reason backward—given the behavior of the graph, the students need to find a possible function that fits that behavior. One possible answer is $g(x) = \frac{1}{x}$. If finding a function is difficult, encourage students to draw one that would fit the requirements.

5. Kathy described a limit using the following statement. What do you think she is describing? Use a graph to represent her statement.

The function $f(x)$ has a limit L as x approaches infinity if, no matter what tiny value we choose, $f(x)$ is closer than this distance to L for some x value and remains closer as x increases, no matter how large x becomes.

This question has students grapple with a "kid-friendly" definition of a limit as x. You may want to challenge students to generate similar statements to make sense of limits at a point or as x approaches negative infinity.

6. Write three different limit statements about the function graphed here.

(See Reproducible 74, Limit Statements from Graph.)

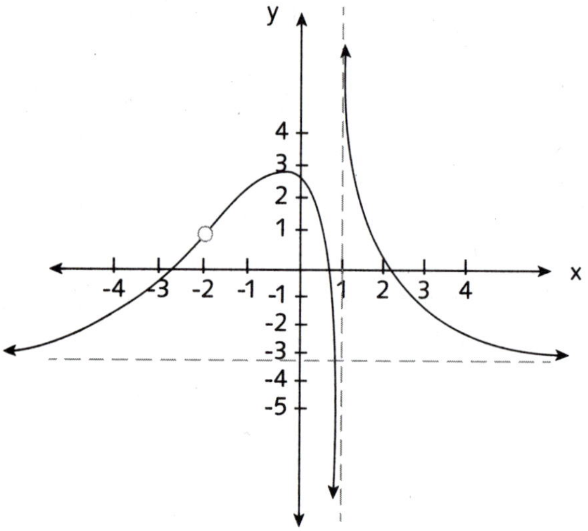

This problem is open-ended and can generate a lot of different limit statements that the class can evaluate. Encourage students to go beyond three limits as others finish. You may want to have students challenge the class with theirs (without giving away the answer). Some teachers provide different conditions—for example, at least one limit must not exist, at least one limit must exist, at least one limit must be one-sided, and so on.

7. Ari is looking at the function $g(x) = 4x + 1$. He claims that the limit of $g(x)$ as x approaches 2 must be 9. He says:

Since $g(x)$ is continuous at $x = 2$, then $g(2)$ exists. Also $\lim_{x\to 2} g(x)$ must exist and these limits are equal to each other. Therefore $\lim_{x\to 2} g(x) = g(2) = 9$.

Do you agree or disagree? Explain your thinking.

Ari has the right answer for the wrong reason. This is an example of circular reasoning. Continuity depends on the limits existing. Therefore, we cannot use continuity to justify that the limits exist. Many students will have no issue with circular logic, so expect a debate. If students are stumped, ask them about the condition of continuity—and what conditions mean logically.

8. When are the following statements equal? When are they not equal? Can only two of them be equal? Explain.

$$\lim_{x\to 2} f(x) \qquad \lim_{x\to 2^-} f(x) \qquad \lim_{x\to 2^+} f(x)$$

This question gets at the role of a two-sided limit. If the left-hand and right-hand limits agree, then the limit as x approaches 2 also agrees. If $\lim_{x\to 2} f(x)$ equals one of the one-sided limits, then it must equal the other. This question can also be adjusted by focusing on existence. Namely, *can only two of them exist? Why or why not?*

9. Review the three conditions of continuity at $x = a$ shown here. Then sketch a function $f(x)$ so that at $(a, f(a))$ the function meets only one condition.

1. f is defined at a
2. $\lim_{x \to a} f(x)$ exists
3. $\lim_{x \to a} f(x) = f(a)$

This type of question requires students to understand the role of each of these conditions to result in continuity. Answers will vary as students may choose either 1 or 2 to focus on. Students should recognize that 3 requires both 1 and 2, and therefore should not be considered. If students focus on meeting only condition 1, they could use a discontinuity at which the point $(a, f(a))$ is defined. If they focus on only meeting condition 2, they may create a hole in an otherwise continuous function.

Area Under the Curve

Experiences at This Level Will Help Students To

- estimate the area under a curve in multiple ways, including with rectangles or trapezoids of equal width
- find exact areas by taking a limit of a Riemann sum as the number of rectangles increases to infinity
- understand that the area of a region below the *x*-axis is negative in order for the area function to be consistent

Reproducibles are available in a downloadable, printable format. See page xx for directions about how to access them.

Materials

- Price of Ride-Share Service Graph (Reproducible 75)

Good Questions and Teacher Notes (pages 294–301)

1. Emily works for a ride-share service for which the price a rider pays is based on the number of minutes the ride takes and the rate for that time of day. The cost per minute is represented in the following graph. Assuming that a person took a ride starting at 2:00 p.m. and ending at 5:00 p.m., approximately how much should they expect to pay (without tip)? Try to make your estimate as accurate as possible.

(See Reproducible 75, Price of Ride-Share Service Graph.)

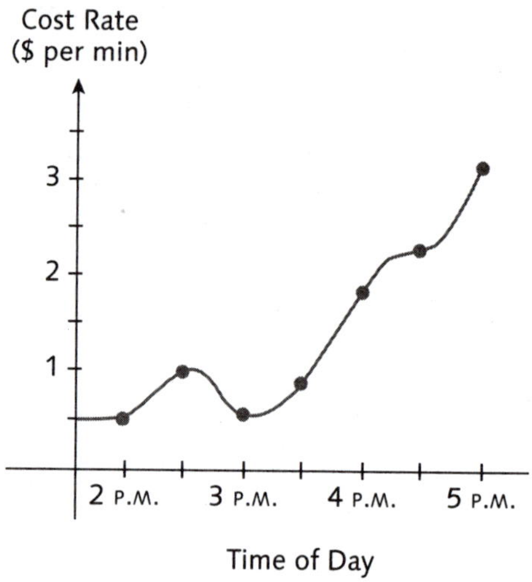

This question gives an opportunity for students to use rectangles or trapezoids or other shapes to dissect the area under the curve to estimate the cost. This question can be asked even if students have never been introduced to any work with area under a curve before. Students who are seeing a question like this for the first time may build a table and use function values to calculate an estimate. If students are not calculating areas, ask "How is your estimate related to the geometry of the curve?" or "How can I represent this cost geometrically?"

2. **Emily works for a ride-share service for which the price a rider pays is based on the number of minutes the ride takes and the rate for that time of day. The cost per minute is represented in the following graph. When could you ride and pay about $75? Identify as many options as you can.**

(See Reproducible 75, **Price of Ride-Share Service Graph.**)

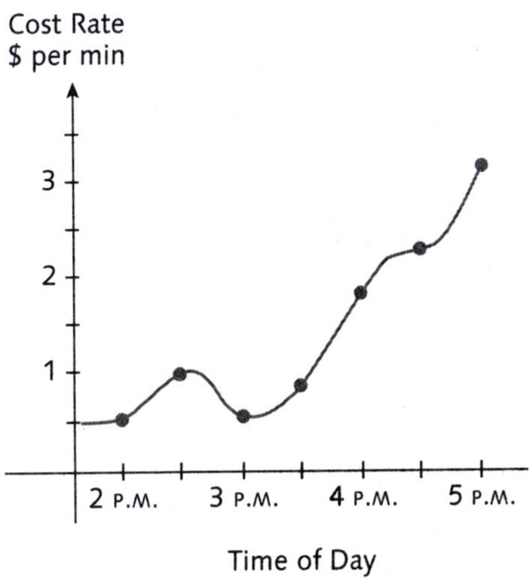

This changes the question slightly and gets students to think in the opposite direction: *what portion of the graph has an area of approximately $75?* Students might think about average rate per hour. That is, if the rate were constant at $1.25 per minute, the cost per hour would be 1.25 × 60 = $75. So students should see that between 2:00 and 3:00 p.m. cannot work, since the average rate between 2:00 and 3:00 p.m. is less than 1. Between 3:00 and 4:00 p.m. is reasonable, and between 4:00 and 5:00 p.m. is not possible. However, for only 30 minutes, the average rate should be double, at $2.50 per minute. Students might propose a 30-minute interval near 4:30 p.m., such as between 4:15 and 4:45 p.m.

3. Emily works for a ride-share service for which the price a rider pays is based on the number of minutes the ride takes and the rate for that time of day. The cost per minute is represented in the following graph. Sketch a new graph that represents the cost of a ride at time x if the ride starts at 2:00 p.m.

(See Reproducible 75, Price of Ride-Share Service Graph.)

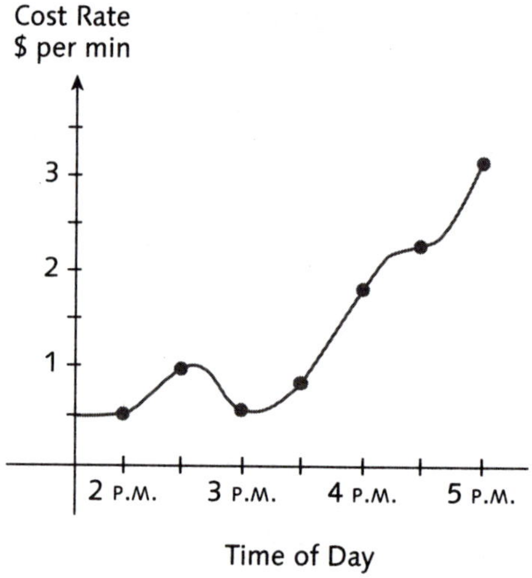

This question focuses on area under the curve as accumulation. Students may benefit from creating a table to accumulate cost for each 30 minutes. At 2:00 p.m., the graph should be on the x-axis. Some students may think the overall cost goes down between 2:30 and 3:00 p.m. because the rate of cost decreases.

4. One focus of calculus is to find the best approximations possible. However, sometimes you already have ways to get exact answers without special calculus tools. With this in mind, find the *exact* area between the curve $y = \sqrt{100 - x^2}$ and the x-axis between $5 \leq x \leq 10$.

This question allows students to use basic geometry. Students can find the area of a sector ($\frac{1}{6}$ of the whole circle with radius 10 units) and then subtract the area of a right triangle with base 5 units and hypotenuse 10 units.

5. Emilio is trying to decide between using left-handed rectangles and right-handed rectangles to approximate the area under the graph of a function. He wants a good estimate without going over. When does using left-hand rectangles result in an underestimate? What does using right-hand rectangles result in an underestimate?

Explain and use a diagram to illustrate your ideas.

Using left-hand rectangles results in an underestimate whenever the curve is increasing and using right-handed rectangles results in an underestimate whenever the curve is decreasing. It will not matter any time the curve is a constant ($y = a$).

6. Josue has a question: "I learned that the exact area under $f(x)$ between a and b can be found by taking the limit of the Riemann sum as the number of rectangles increases to infinity. However, since the width of the rectangles goes to 0, then why isn't the area 0?" How would you respond?

$$\lim_{n\to\infty}\sum_{i=1}^{n} f\left(a + i\left(\frac{b-a}{n}\right)\right)\left(\frac{b-a}{n}\right)$$

Expect this question to spark an interesting debate among students. This question evokes a conundrum that requires students to think about infinitesimals and to revisit the meaning of limits. Josue's challenge is that he is evaluating a limit by trying to calculate the Riemann sum expression for infinity. However, since infinity is not finite, it does not exist and this thinking is problematic. Instead, we have to view the number of rectangles as increasing unbounded, so their widths continue to shrink but are not 0.

7. Explain what the integral here represents. Sketch a possible graph with $f(x)$ and $g(x)$ that shows your thinking.

$$\int_a^b (f(x) - g(x))\,dx$$

This question is appropriate before the area between two curves is introduced to students. If students have no experience of interpreting integrals, you may want to ask questions like, *If this were a Riemann sum, what would be the height of the rectangles? What would be the width?*

8. Compare and contrast the following expressions. What is the same and/or different? What statements can you say about these expressions? Use a graph to demonstrate your thinking.

$$\int_a^b (f(x) - g(x))\,dx \qquad \int_a^b (g(x) - f(x))\,dx$$

Building on the interpretation of integrals, this question directs students to think about the importance of the order of the functions in the expression. That is, we want students to realize that $f(x) - g(x)$ is not the same as $g(x) - f(x)$ and that $\int_a^b (f(x) - g(x))\,dx = -\int_a^b g(x)\,dx + \int_a^b f(x)\,dx$. Students may assume that these integrals are equal.

9. How are the integrals below the same or different? How are they related? Use a graph to show the relationships.

$$\int_3^8 2x\,dx \qquad \int_3^8 (2x + 7)\,dx \qquad \int_3^8 7\,dx$$

This question prompts a useful property of integrals, namely $\int_a^b (f(x) + g(x))\,dx = \int_a^b f(x)\,dx + \int_a^b g(x)\,dx$. Students can develop geometric arguments, showing that the region under the curve $f(x) + g(x)$ can be subdivided into parts representing $\int_a^b f(x)\,dx$ and $\int_a^b g(x)\,dx$.

10. Complete the table shown for the area function $A(x)$, which represents the area under $f(x) = -2x + 10$ between 0 and x. Then determine $A(x)$. Based on $A(x)$, for what x is $A(x) = 0$? How does this connect with the graph of $f(x)$?

x	Area under $f(x)$ between 0 and x
0	
1	
2	
3	
⋮	

This problem sets up the opportunity to see why the area that is under the x-axis is negative. Instead of memorizing this fact, students can see that it needs to be negative for consistency of the system. That is, since $A(x) = -x^2 + 10x = x(10 - x)$, then $A(x) = 0$ when $x = 0$ and when $x = 10$. This connects with the graph two different ways. When $x = 0$, then there is no region between 0 and 0, so the area is 0. But when x is 10, then for the total area between 0 and 10 to be 0, the area under the x-axis must be negative. Note that this question can be asked early on in the study of calculus. Students can determine the area function multiple ways, but most will find patterns in the table. For example, some students may know how to use second differences to know it's quadratic. Other students might see the pattern that $A(1) = 1 \times 9$, $A(2) = 2 \times 8$, $A(3) = 3 \times 7$, and so forth.

11. When $f(x) = -2x + 10$, the area function $A(x) = -x^2 + 10x$, which represents the area under $f(x)$ between 0 and x. Based on this area function, what is $A(-2)$? How does this make sense graphically?

This question continues the thinking generated from Question 9 to have students recognize why the area above the x-axis, with bounds that are considered from right to left, must be negative. That is because $A(-2) = -(-2)^2 + 10(-2) = -24$. You may want to follow up by asking students to predict what would happen to area from right to left for a region below the x-axis.

Slope of Tangents, Derivatives

Experiences at This Level Will Help Students To

- understand that derivatives at a point $f'(a)$ represent the slope of the tangent to the graph of a function at that point
- interpret the notation $f'(x)$ as a representation of a derivative function, where the output for all $x = a$ is the slope of the tangent to the graph of $f(x)$ at a
- accept multiple definitions of a derivative, and the idea that each can be thought of geometrically in relation to the slope of a secant

Reproducibles are available in a downloadable, printable format. See page xx for directions about how to access them.

Materials

- Aria's Notes (Reproducible 76)
- Arthur's Graph Design (Reproducible 77)
- Roberto's Graph (Reproducible 78)
- Venn Diagram with Curve Features (Reproducible 79)
- Keniesha's Graph (Reproducible 80)

Good Questions and Teacher Notes (pages 303–310)

1. Kayla is trying to estimate the velocity of an object traveling with a displacement over time of $f(t)$ at $t = 3$ seconds. She is trying to decide between finding the slope of the two secants as shown here. What is the same or different between these secants? And assuming h is small, which do you think she should select and why?

$$\frac{f(3+h) - f(3-h)}{2h}$$

$$\frac{f(3+h) - f(3)}{h}$$

This problem requires students to connect the slope expressions often used in the definition of a derivative with its geometric interpretation. The limit of the slope as h approaches 0 is not given because, if so, the limits should be equal. However, for small values of h, students should recognize that the top expression calculates an average change across $3 - h \leq x \leq 3 + h$ while the bottom expression is the slope of a secant on only one side of $f(3)$ over $3 \leq x \leq 3 + h$. Even though h is small, these will likely offer different estimates of the slope unless $f(t)$ is constant. Yet students may recognize that the top expression represents an average velocity over an interval upon which 3 is a center, and thus may represent a better estimate.

2. Compare and contrast these definitions of the derivative. How are they the same or different? When might you use each?

$$\lim_{x \to a} \frac{f(x) - f(a)}{x - a} \quad \text{and} \quad \lim_{h \to 0} \frac{f(x + h) - f(x)}{h}$$

Answers will vary. The point of this problem is to have students consider the practical aspects of these different derivative definitions. Some students might recognize that determining $f(x + h)$ is sometimes challenging, yet others might note that dividing by h (instead of $x - a$) is easier. You might want to suggest that students try both approaches for a few different functions, such as $f(x) = x^5$ or $f(x) = \frac{1}{x}$.

3. Aria took notes about her travel home from school. Draw a possible distance-time graph that could represent her movement. Is there more than one?

(See Reproducible 76, Aria's Notes.)

- *I had to double back at least once during my trip home.*
- *I was traveling faster at the beginning of my trip, but I got tired and slowed down.*
- *The trip took 30 minutes.*
- *I live 1 mile from school.*

This problem has students use information about speed and distance to construct a potential distance-time graph. Students should recognize that the graph should be steeper for times closer to 0 and less steep overall for times closer to 30 minutes. Ask students about the speeds when Aria turned around.

4. Arthur made a design with his graph shown here. His teammate Jake said, "Arthur, I think your design can tell us about the changing slope." What do you think he means? Make as many observations about how the slope changes as you can.

(See Reproducible 77, Arthur's Graph Design.)

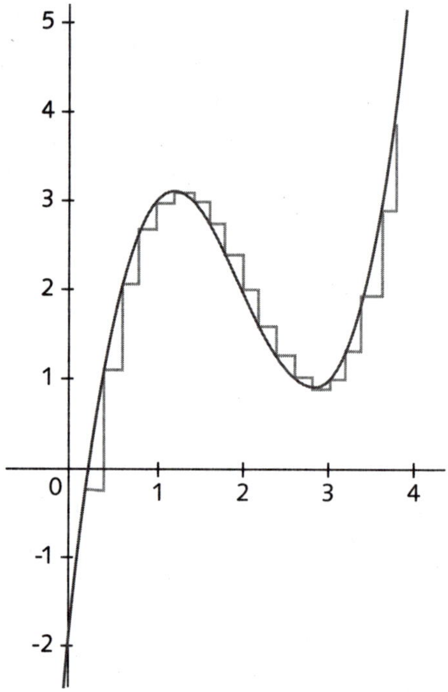

Some students find it difficult to visualize how the slope of a tangent dynamically changes as x changes. This problem offers these students a new way to perceive how the slope changes as x changes. Expect students to focus on the constant change in x for each of these "stair steps" but that the different changes in y for each represents a different amount of change in the slope.

5. Roberto zoomed in on a graph on his computer, and the result is shown here. What could be his function? What could it not be? Explain.

(See Reproducible 78, Roberto's Graph.)

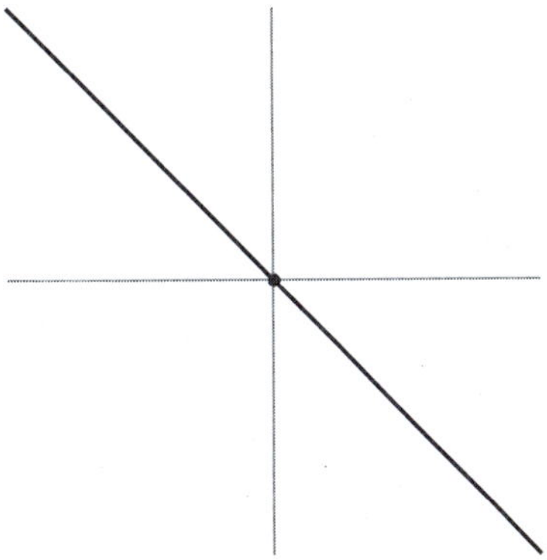

This question gets at the concept of local linearity. That is, when zoomed in far enough, a function that is not linear can appear linear. Expect students to consider functions such as $y = -\sin(x)$ or $y = -x^3$ as potential functions. Have students also contribute why it could not be others, and expect reasons around slope or y-intercepts at $x = 0$.

6. Draw the graph of four different functions, one that satisfies each region of the Venn Diagram at $x = 2$. Explain how the first and second derivative at $x = 2$ is the same or different for each function.

(See Reproducible 79, Venn Diagram with Curve Features.)

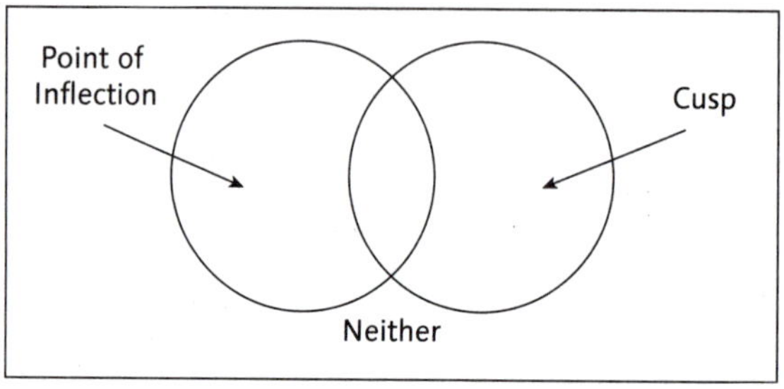

This task gets at the variations of ways the second derivative controls the features of the graph (i.e., whether there is a cusp, point of inflection, both, or neither). This question works well after students have been introduced to the second derivative and points of inflection, but who still need to make sense of the relationship between the curvature and the second derivative.

7. If $f'(x)$ is an even function, and if $f(0) = 0$, what can you say about $f(x)$? Is $f(x)$ even, odd, or neither? What if $f'(x)$ is an odd function? Explain your reasoning.

This question requires students to think about symmetry and its relationship between a function and its derivative. When the first derivative is an even function, this means that the slopes of $f(x)$ must be symmetric. Ask students what this tells them about the shape of $f(x)$. Since $f(0) = 0$, this means that $f(x)$ is an odd function. Students can test this conclusion using even functions they know well, such as $y = \cos(x)$ and $y = x^2$.

8. Keniesha knows that the graph of $f(x)$ has a cusp. However, she cannot remember if the following graph is $f(x)$ or $f'(x)$. Which do you think this graph represents? Could it be either? Justify your conclusions.

(See Reproducible 80, Keniesha's Graph.)

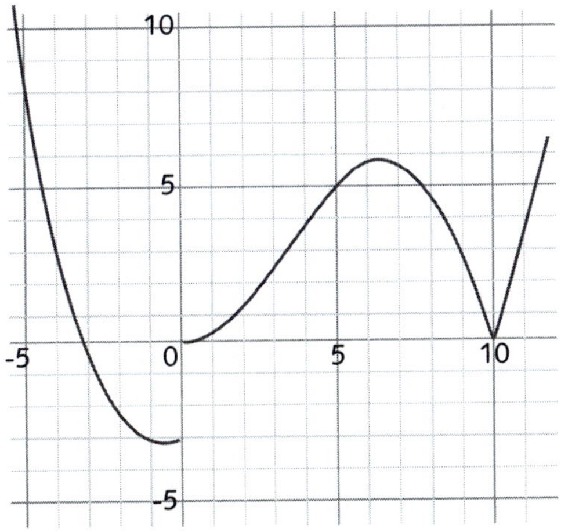

This question requires students to think about the conditions of the slope of a function at a cusp. This graph has a cusp so it can certainly be $f(x)$. However, it could also be $f'(x)$ since there is a discontinuity at $x = 0$. If this graph is $f'(x)$, then the slope when $x < 0$ is negative and the slope when $x > 0$ is positive. Have students explain that if the limit of the slope as $x \to 0^-$ does not equal the limit of the slope as $x \to 0^+$ then a cusp is formed.

9. Use two different strategies to find the derivative of $h(x) = \frac{f(x)}{g(x)}$. Be ready to explain your work.

Students who are aware of the quotient rule can use that to differentiate $h(x)$. However, when looking for a second way, students may think to rewrite the equation as $h(x)g(x) = f(x)$ and use implicit differentiation. Or they could rewrite the equation as $h(x) = f(x)(g(x))^{-1}$ and use the product rule.

10. Draw a graph of $f(x)$ that has the features described below. Is there more than one possible solution? Explain.

- $f'(x) > 2$ for all $x > -1$ and $f'(x) < -1$ for all $x < -1$.
- $f(x) = 0$ has solutions at $x = -4$ and $x = 0$.

This question challenges students to coordinate information about first derivatives and the function to draw a potential graph. Students should figure out that $f(x)$ must have a cusp at $x = -1$.

Fundamental Theorem of Calculus

Experiences at This Level Will Help Students To

- make connections between the derivative of an integral and the rate of change of an area function
- recognize why the derivative of an antiderivative of $f(x)$ is $f(x)$, while the antiderivative of the derivative of $f(x)$ is $f(x) + C$

Reproducibles are available in a downloadable, printable format. See page xx for directions about how to access them.

Materials

- Motion Detector Graph (Reproducible 81)

Good Questions and Teacher Notes (pages 312–315)

1. How is the area under the curve of $f(x) = -4x + 12$ changing? Calculate the area under the curve (called $A(x)$) between 0 and x for $0 \leq x \leq 5$ and write the equation for $A(x)$. How does $A(x)$ change? How is the change related to $f(x)$?

This question prompts students to recognize the relationship between the change of the area function (which behaves similar to the derivative of an integral) and the function itself. Students should start by calculating $A(0) = 0$, $A(1) = 10$, $A(2) = 16$, $A(3) = 18$, $A(4) = 16$, and $A(5) = 10$ and generalizing this with the equation $A(x) = x(12 - 2x)$. Students can reason about this geometrically (using a generic trapezoid with bases 12 and $12 - 4x$ lengths and height of x), using constant differences, or by guessing and checking.

2. What is similar and different about the two functions described here?

The area function under	The change function for
$f(x) = a$	$g(x) = ax$

This question also has students recognize that the area function under $f(x) = a$ is $A(x)= ax = g(x)$, and that the change function for $g(x) = ax = a = f(x)$. You may want to ask students what other functions could be $f(x)$ that would keep them the same. You might start by asking if $f(x) = ax$, then what does $g(x)$ need to be?

3. Kirsten has a motion detector that measures your distance from a wall as you move. It gives you two graphs, one for your distance from the wall over time and the other for your velocity over time. The two graphs are shown here, but unfortunately all labels are erased. Which is which? How do you know?

(See Reproducible 81, Motion Detector Graph.)

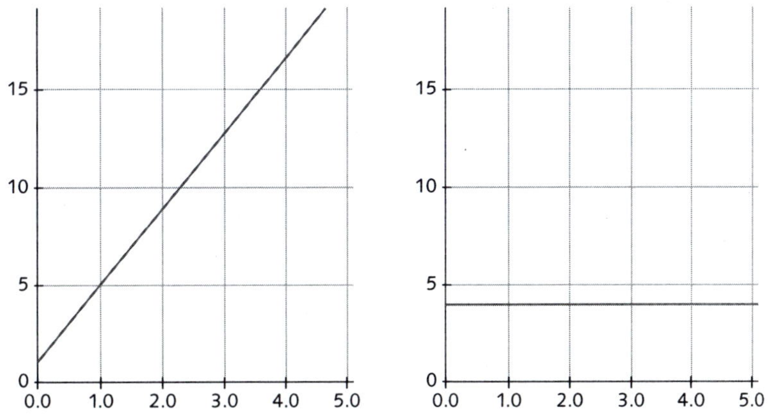

This question challenges students to make sense of the connections between these two data sets. If the left-hand graph represents distance, then the velocity is its slope, which is constant. Therefore, the right-hand graph is the velocity. However, it is reasonable to ask students what the graph of the velocity would look like if the right-hand graph were the distance function. Challenge students to make connections between these two functions by asking questions such as *The right-hand graph represents* $y = 4$. *How does that relate to the function in the left-hand graph? Explain your thinking.*

4. If $A(x) = \int_0^x (7 - 2t)\,dt$ and $B(x) = \int_3^x (7 - 2t)\,dt$, then what is the relationship between $A'(x)$ and $B'(x)$? Explain.

This question can help students think about how the derivative of an infinite number of antiderivatives that differ only by a constant will be equal. You may want to ask, *How could the answer to B(x) be rewritten in terms of A(x)?* During a discussion of this question, students should connect their reasoning for this question to the role of the +*C* when determining the antiderivative of a function.

5. What are statements you can make about the expression here? What is its value as h approaches 0? Explain your thinking.

$$\frac{\int_a^{x+h} f(t)\,dt - \int_a^x f(t)\,dt}{h}$$

In a discussion of this question, expect students to recognize the definition of a derivative. Namely, the limit of this expression as h approaches 0 is analogous to $\lim_{h \to 0} \frac{f(x+h) - f(x)}{h} = f'(x)$. Therefore, the limit of this expression is the derivative of $\int_a^x f(t)\,dt$ with respect to x, which represents how the area under $f(t)$ changes as the upper bound (x) changes. The fundamental theorem of calculus states that this is $f(x)$.

6. How are the expressions below the same or different? Explain.

$$\frac{d}{dx}\int_2^x (5t-3)\,dt \qquad \frac{d}{dx}\int_2^9 (5t-3)\,dt \qquad \int_2^x \frac{d}{dt}(5t-3)\,dt \qquad \int_x^2 \frac{d}{dt}(5t-3)\,dt$$

These four expressions represent a range of different conditions of the fundamental theorem of calculus. Students can contrast pairs of them. For example, $\frac{d}{dx}\int_2^x (5t-3)\,dt$ and $\frac{d}{dx}\int_2^9 (5t-3)\,dt$ are different because $\int_2^x (5t-3)\,dt$ is a function of x while $\int_2^9 (5t-3)\,dt$ is a constant. So, $\frac{d}{dx}\int_2^9 (5t-3)\,dt = 0$ while $\frac{d}{dx}\int_2^x (5t-3)\,dt = 5x-3$. It is also worth having students focus attention on what happens with constants and why.

Reproducibles

The following reproducibles are referenced throughout the text. These reproducibles are also available in a downloadable, printable format. For access, visit www.mathsolutions.com/myonlineresources and register your product using the key code **GQHS**. See (Roman numeral) page xx for more detailed instructions.

(Continued)

(Continued)

REPRODUCIBLE 1

Equations Venn Diagram

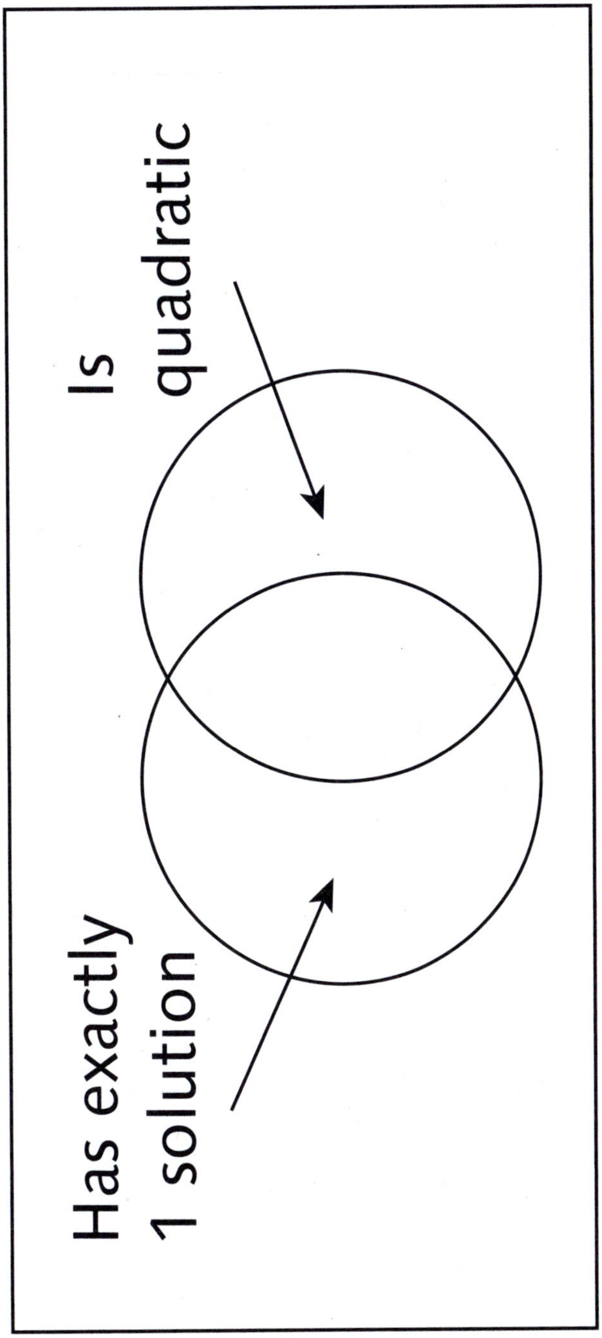

For use with Chapter 5, Question 11, section Solving Equations.

Reproducible 2

Area Diagram A

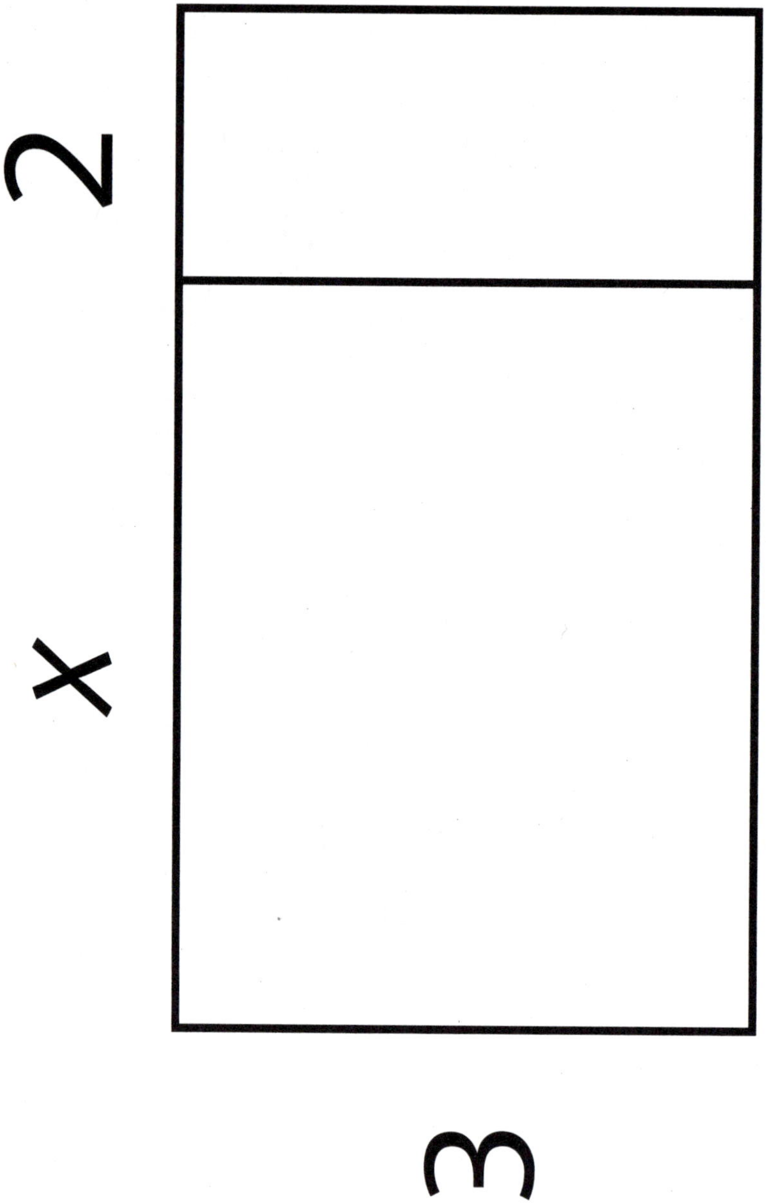

For use with Chapter 5, Question 2, section Equivalence.

Area Diagram B

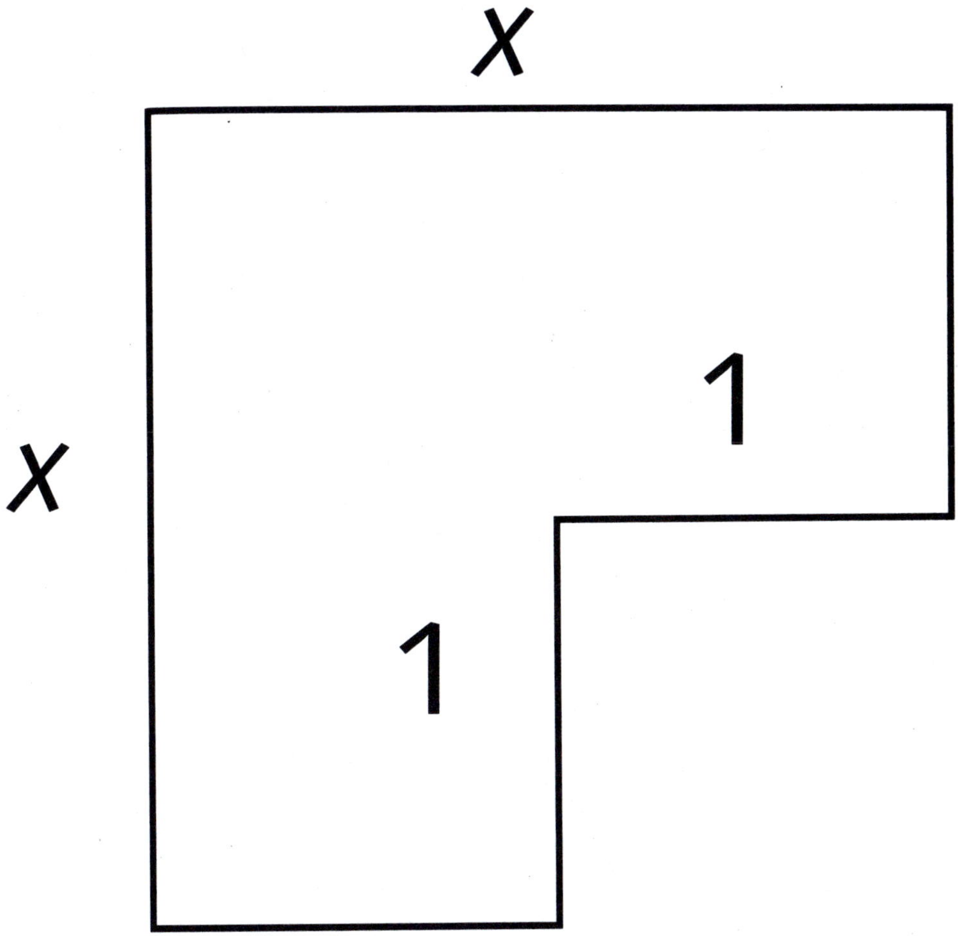

For use with Chapter 5, Question 4, section Equivalence.

Reproducible 4

Graph of Expressions *A* and *B*

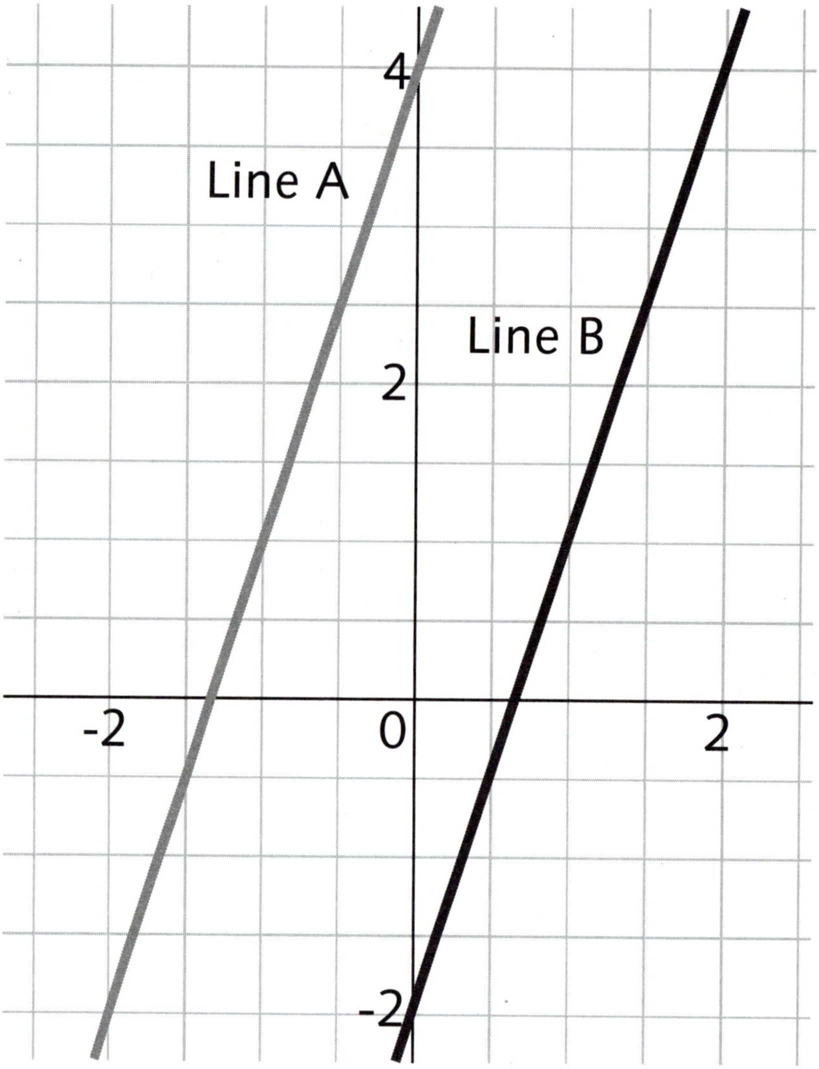

For use with Chapter 5, Question 6, section Equivalence.

From *Good Questions for Math Teaching, High School* by Nancy Anderson, Leslie Dietiker, and Gregg Reilly. Visit www.mathsolutions.com/myonlineresources to register your resource and download the reproducibles. See page xx for the key code.

REPRODUCIBLE 5

Points (x, y) Venn Diagram

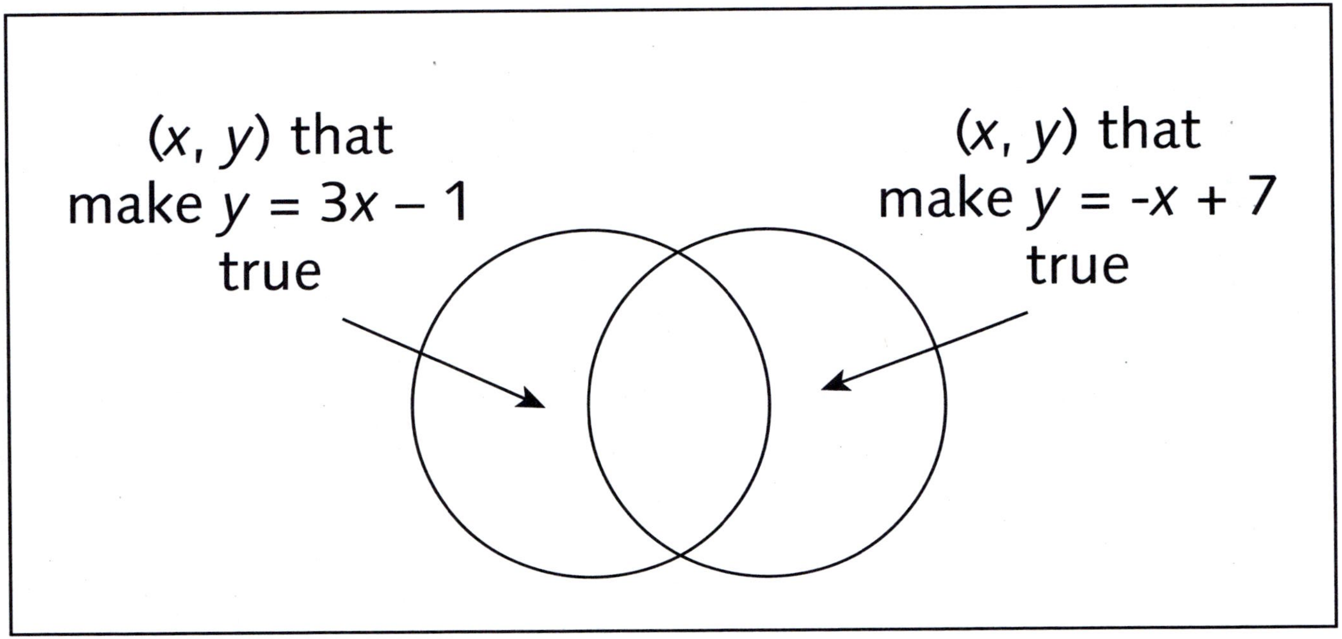

For use with Chapter 5, Question 1, section Systems.

Reproducible 6

Annie's Graph

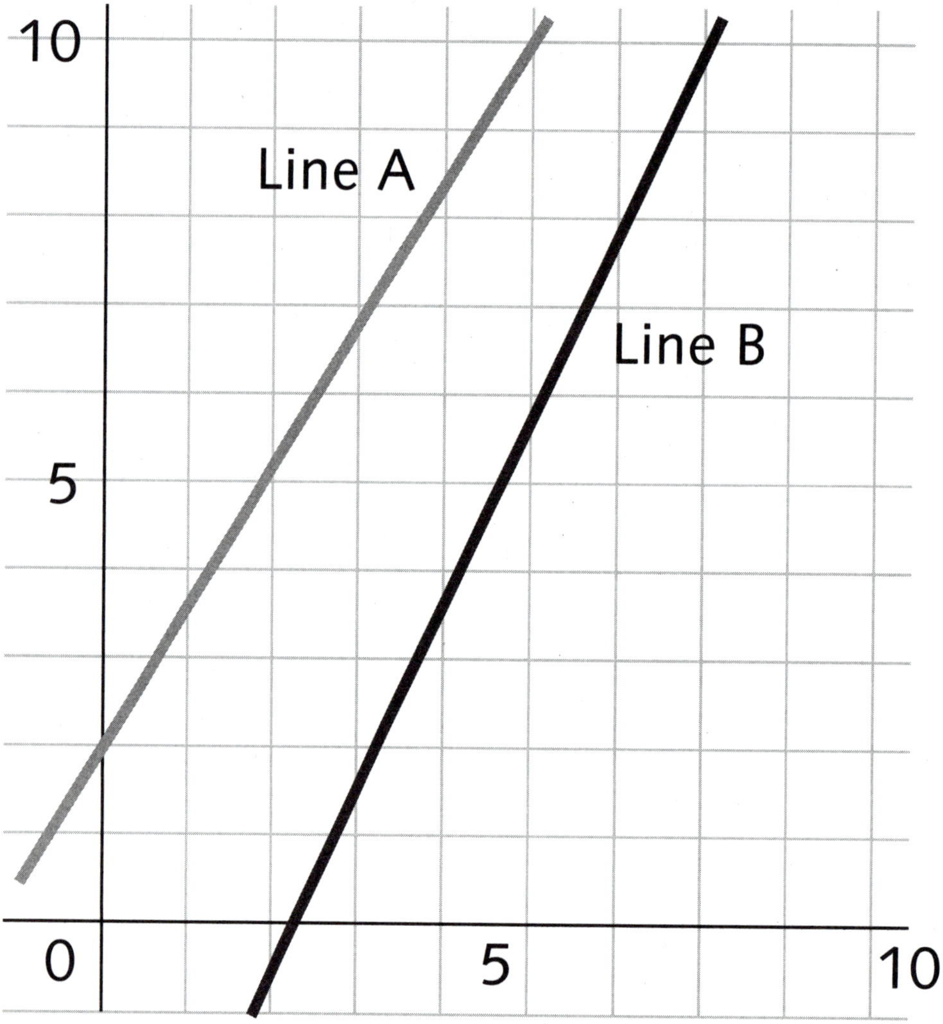

For use with Chapter 5, Question 4, section Systems.

Reproducible 7

Graph of Transformed Functions

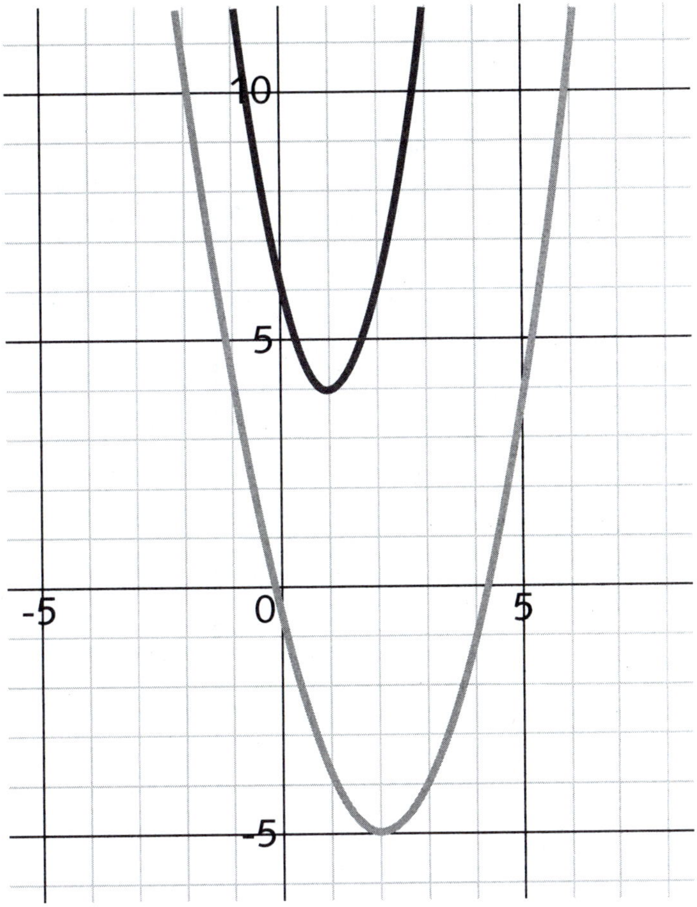

$$f(x) = 2x^2 - 4x + 6$$

$$g(x) = x^2 - 4x - 1$$

For use with Chapter 5, Question 9, section Systems.

Reproducible 8

Area of Rectangle

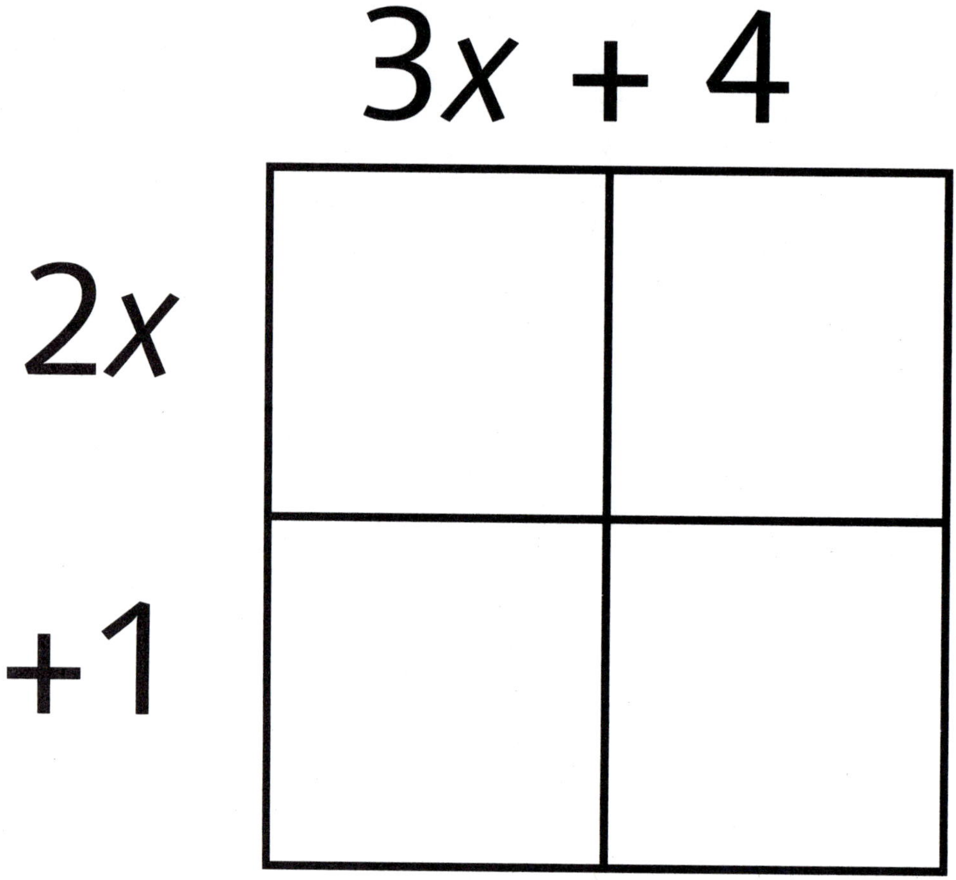

For use with Chapter 5, Question 1, section Rewriting Expressions.

Reproducible 9

Expressions Venn Diagram

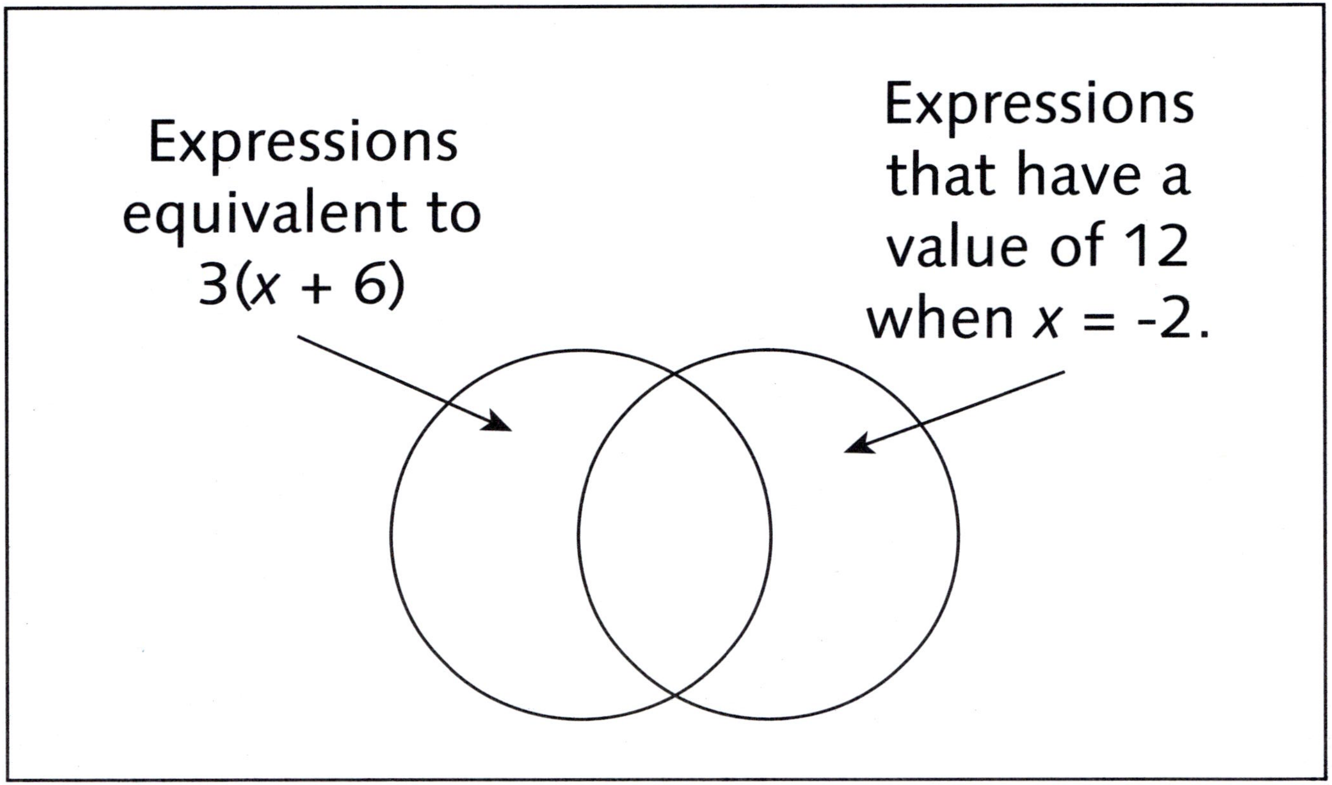

For use with Chapter 5, Question 8, section Rewriting Expressions.

REPRODUCIBLE 10

Solutions Venn Diagram

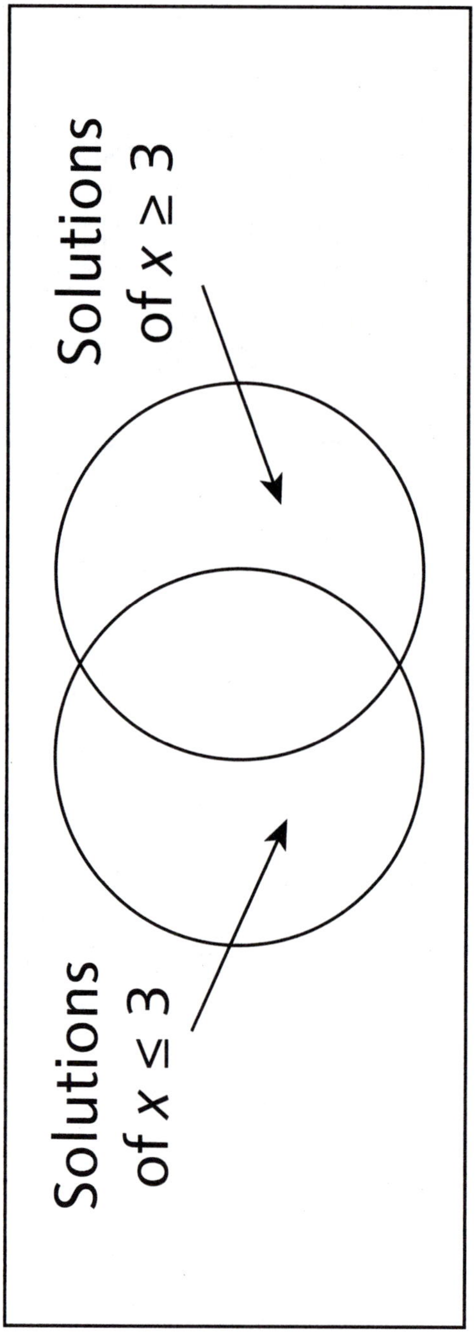

For use with Chapter 5, Question 3, section Solving Inequalities.

Reproducible 11

Gerald's Graph

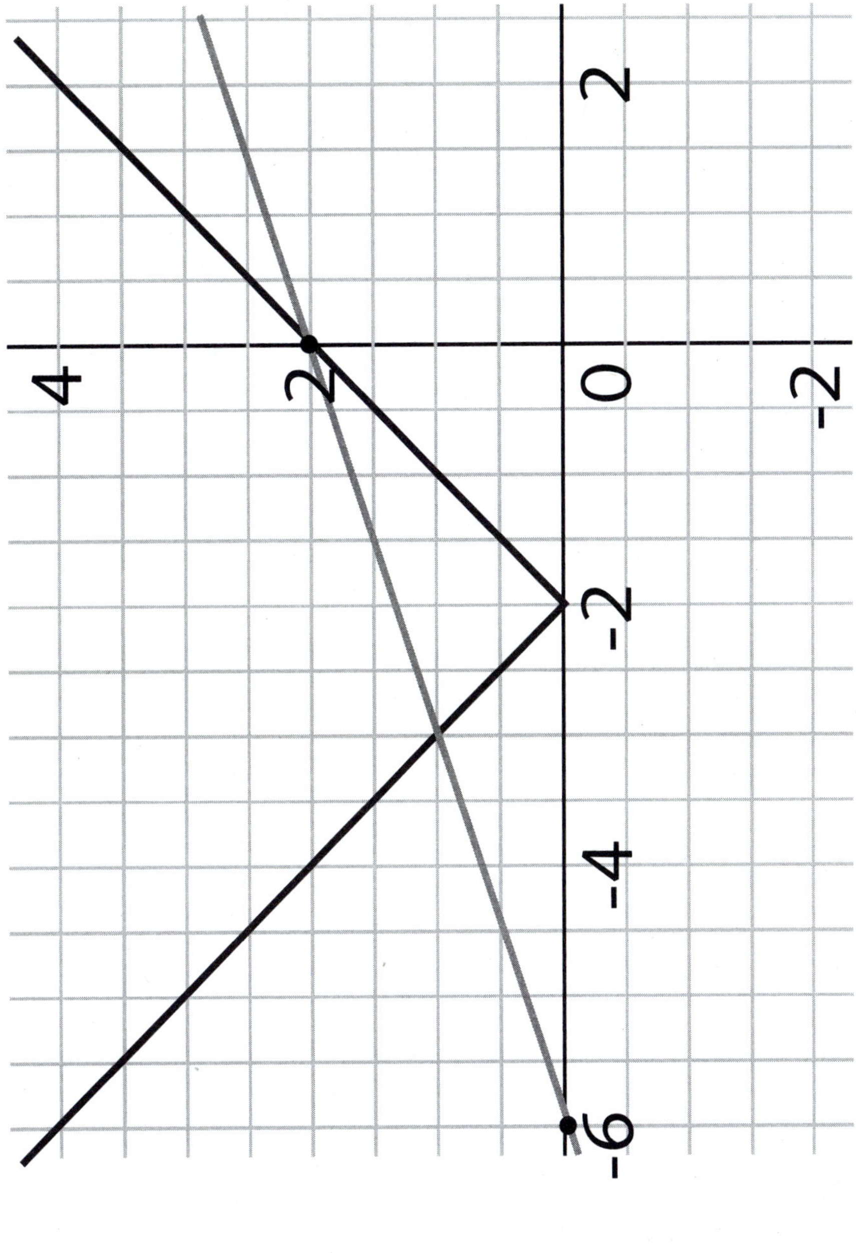

For use with Chapter 5, Question 8, section Solving Inequalities.

REPRODUCIBLE 12

Inequalities Venn Diagram

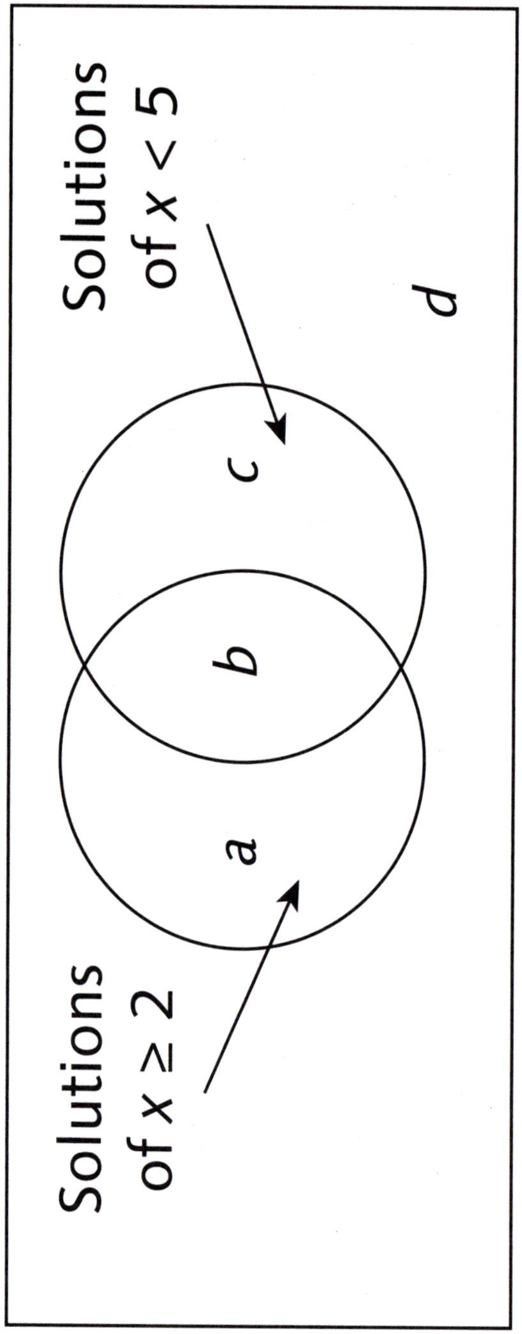

For use with Chapter 5, Question 9, section Solving Inequalities.

Reproducible 13

AB's Graph

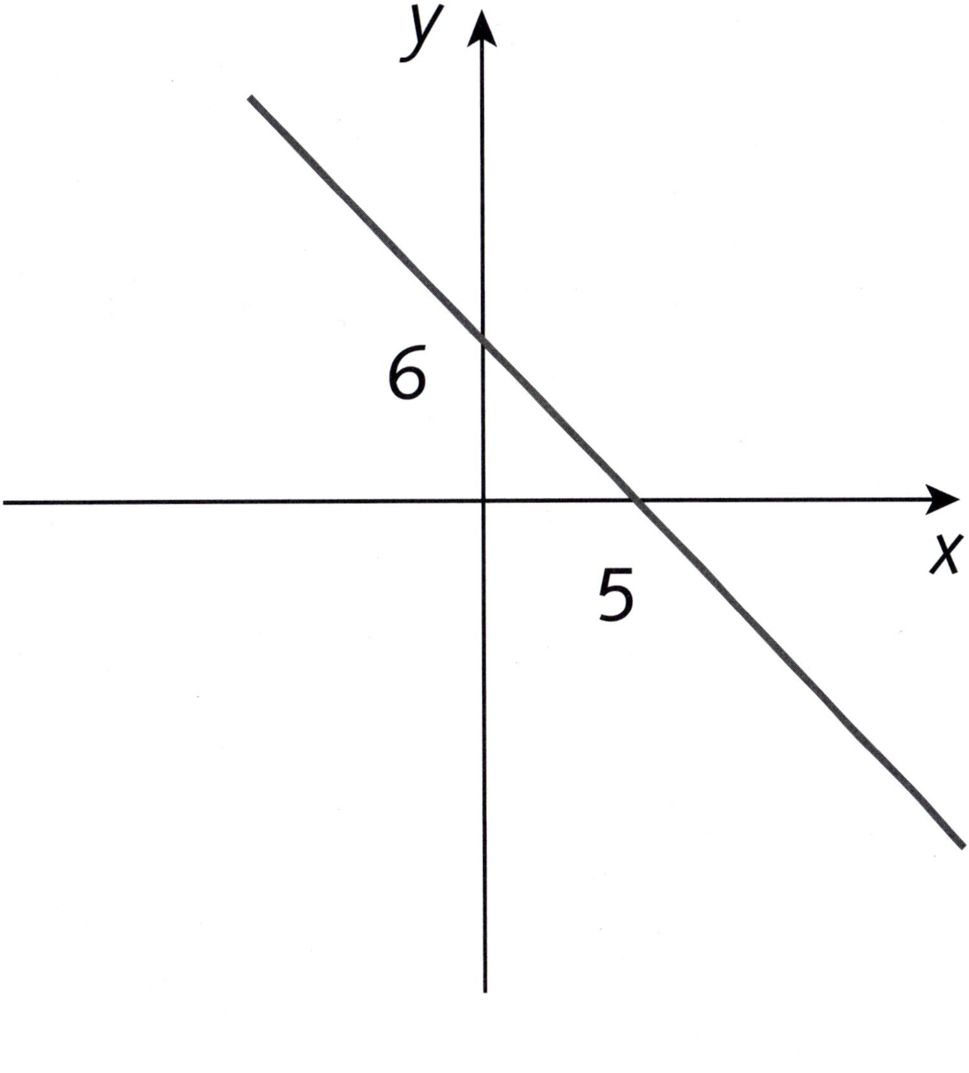

For use with Chapter 6, Question 3, section Linear Functions.

Reproducible 14

Mystery Values and Graph

x	y
-4	-2
-1	-5
5	-11
8	-14

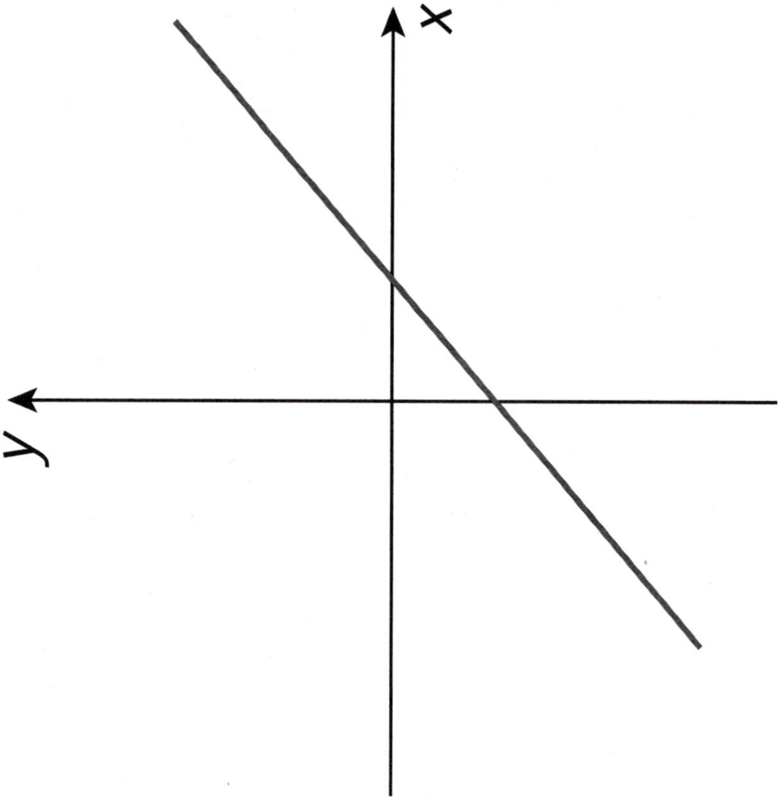

For use with Chapter 6, Question 14, section Linear Functions.

Reproducible 15

Graph of $f(x) = x^2$ and Mystery $h(x)$

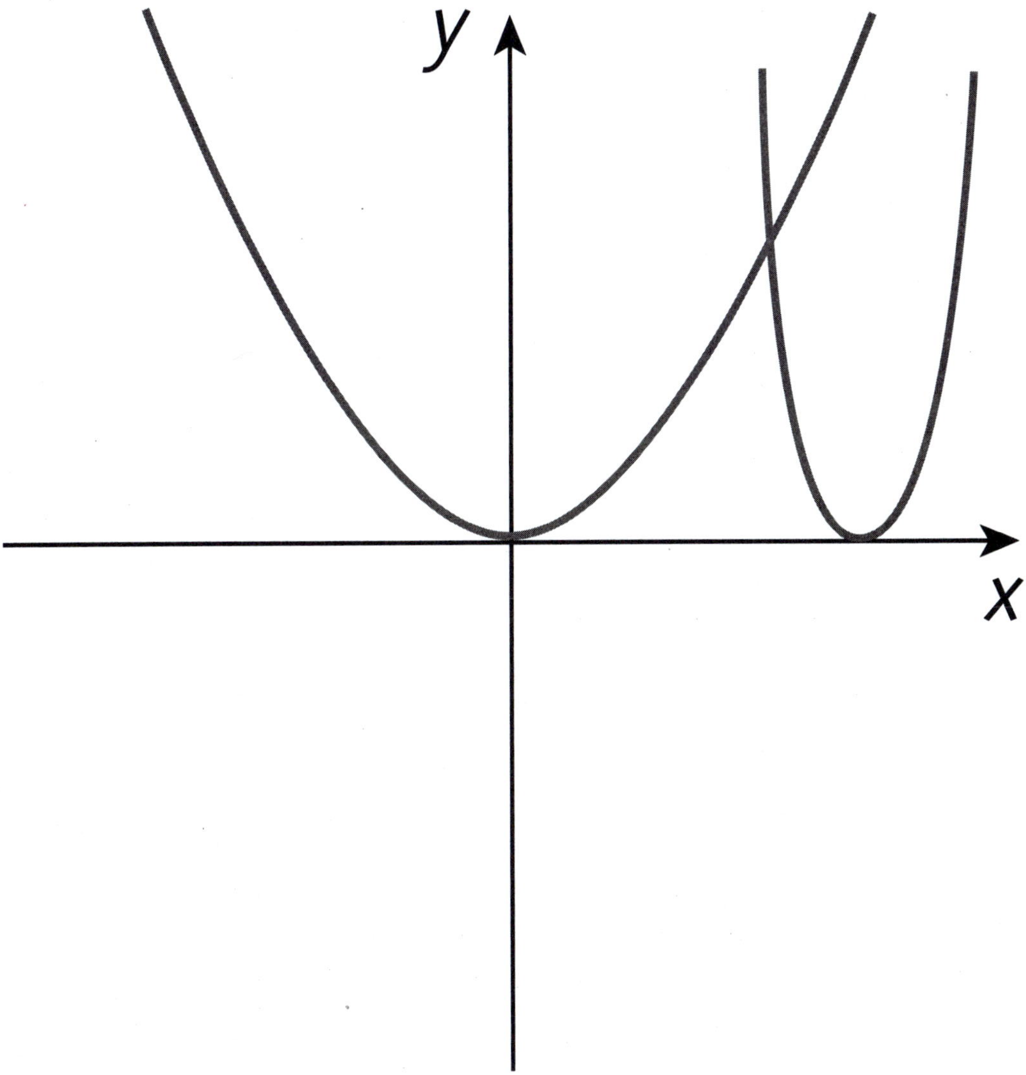

For use with Chapter 6, Question 13, section Quadratic Functions.

Reproducible 16

Graph of Mystery Quadratic Function

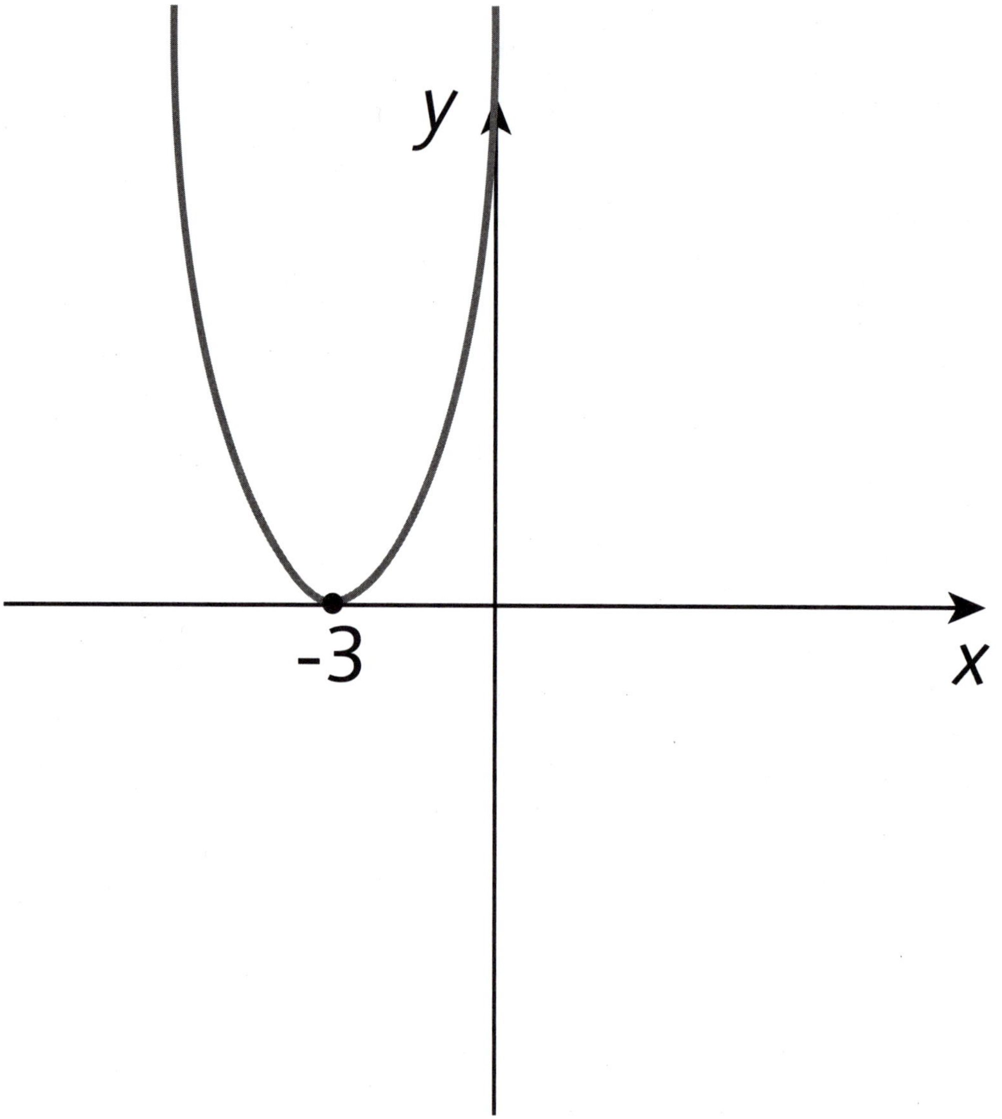

For use with Chapter 6, Question 14, section Quadratic Functions.

Reproducible 17

Graph of Translation of $g(x) = 2x^2$

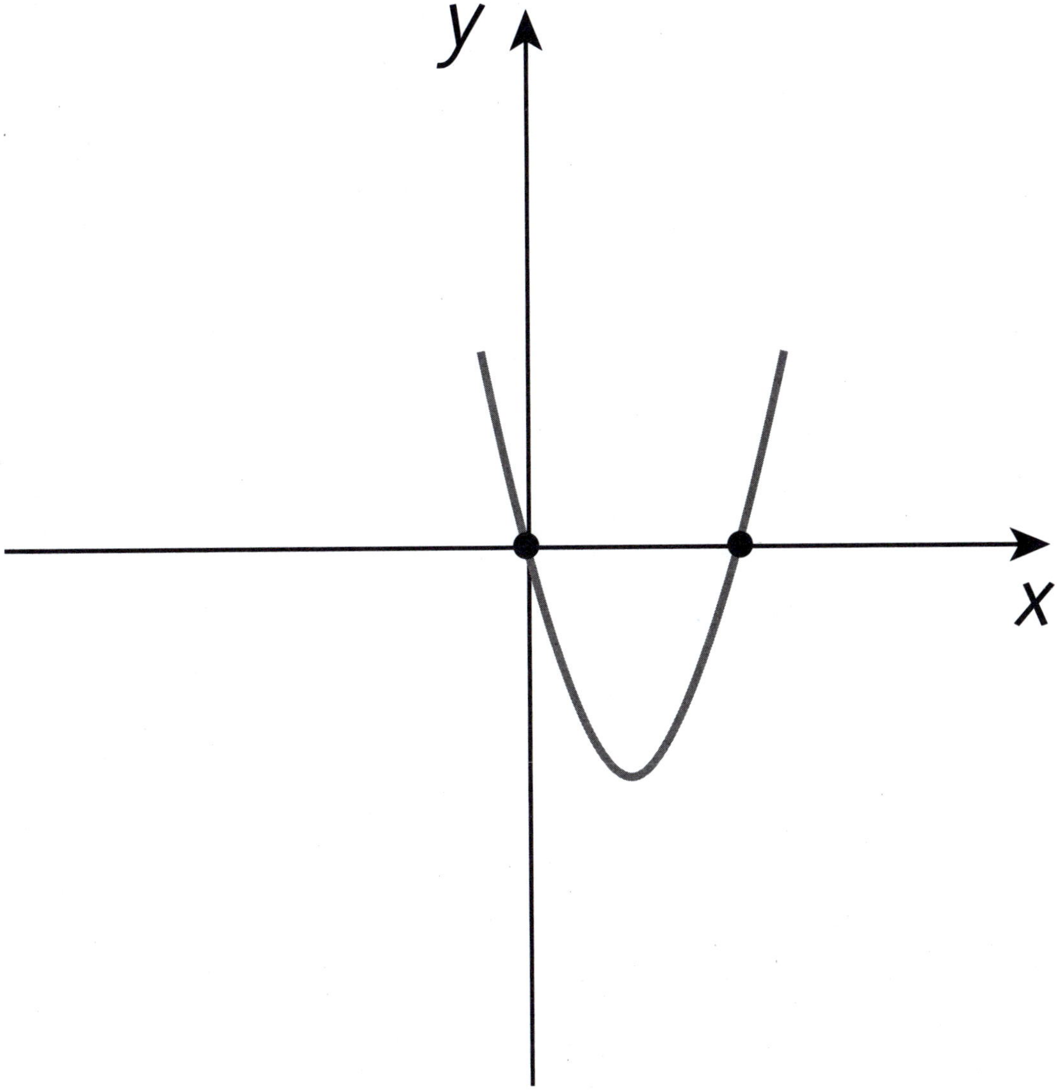

For use with Chapter 6, Question 15, section Quadratic Functions.

Reproducible 18

GC's Graph

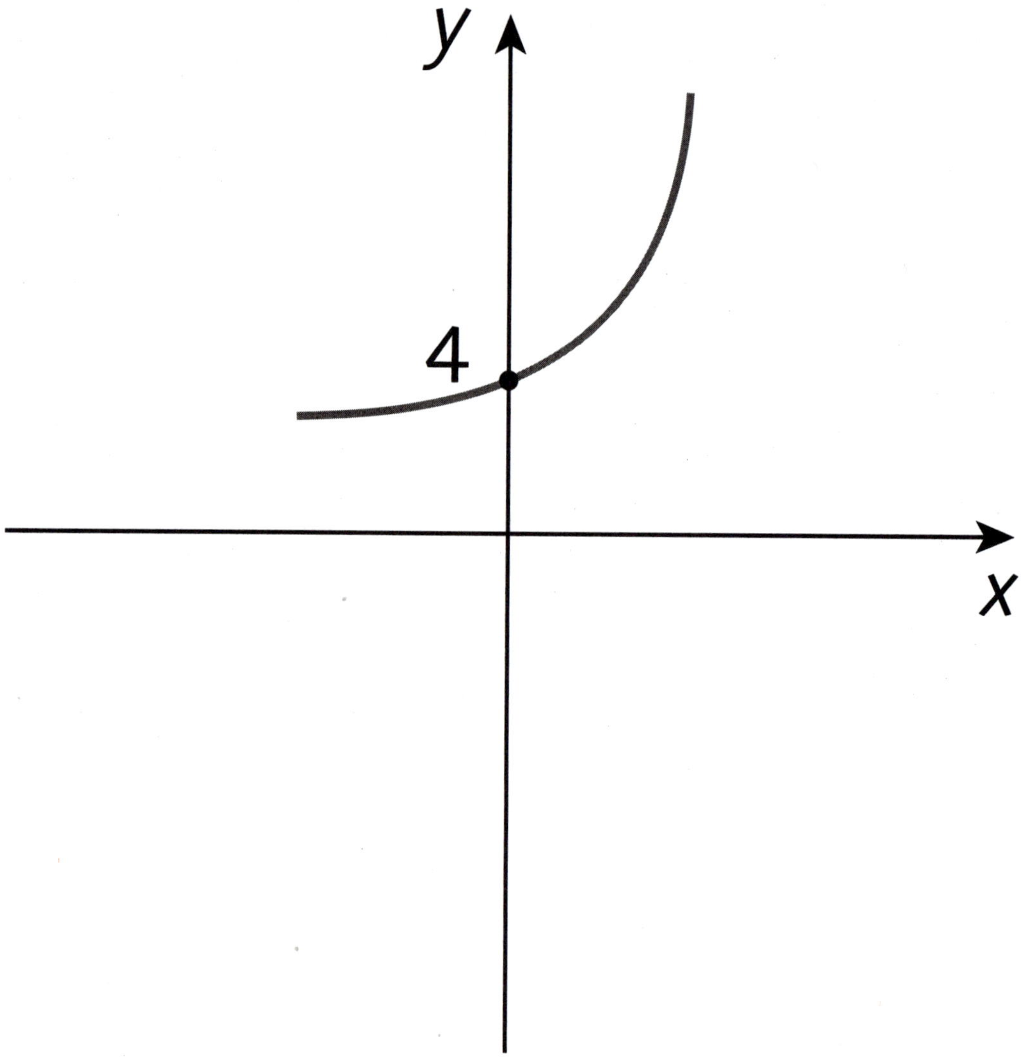

For use with Chapter 6, Question 1, section Exponential and Logarithmic Functions.

Reproducible 19

Graph of Mystery Exponential Function

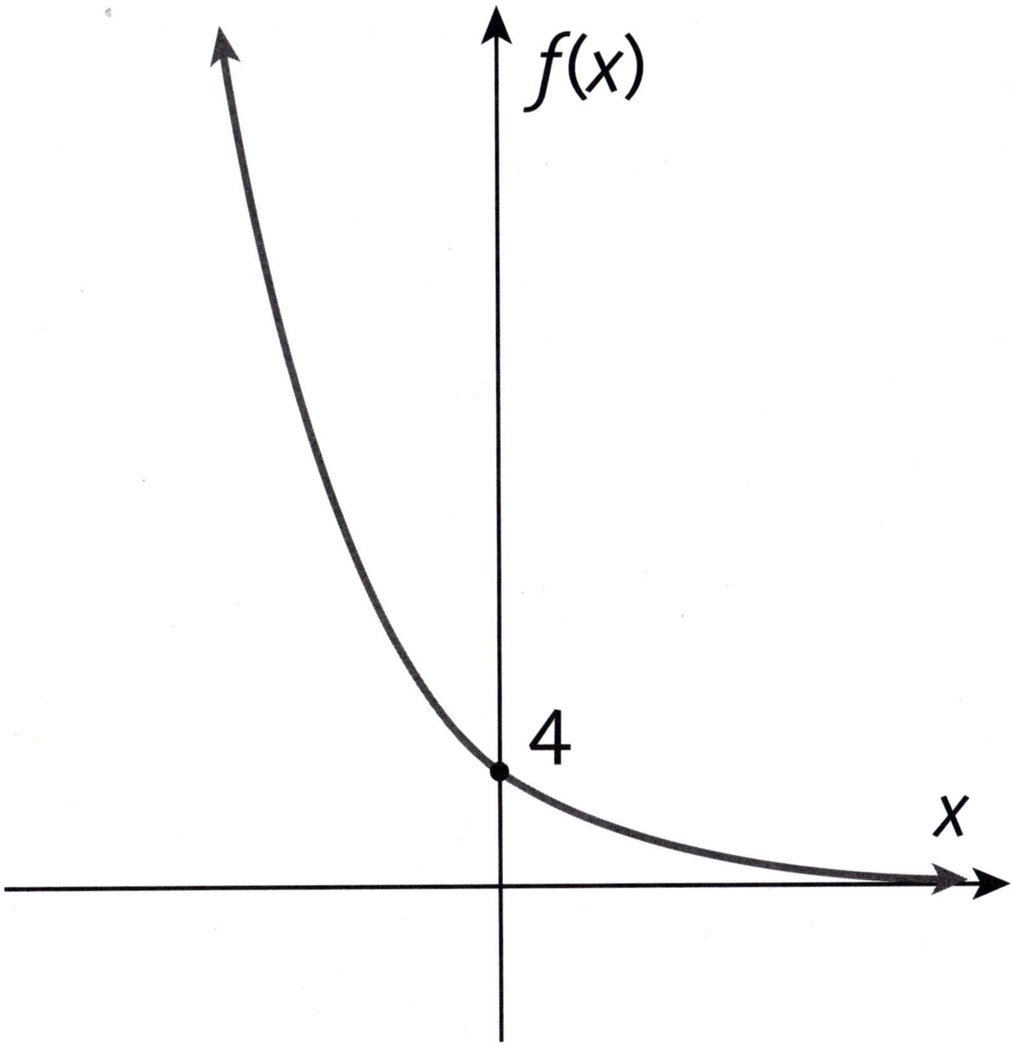

For use with Chapter 6, Question 2, section Exponential and Logarithmic Functions.

Reproducible 20

Graphs of Two Mystery Exponential Functions

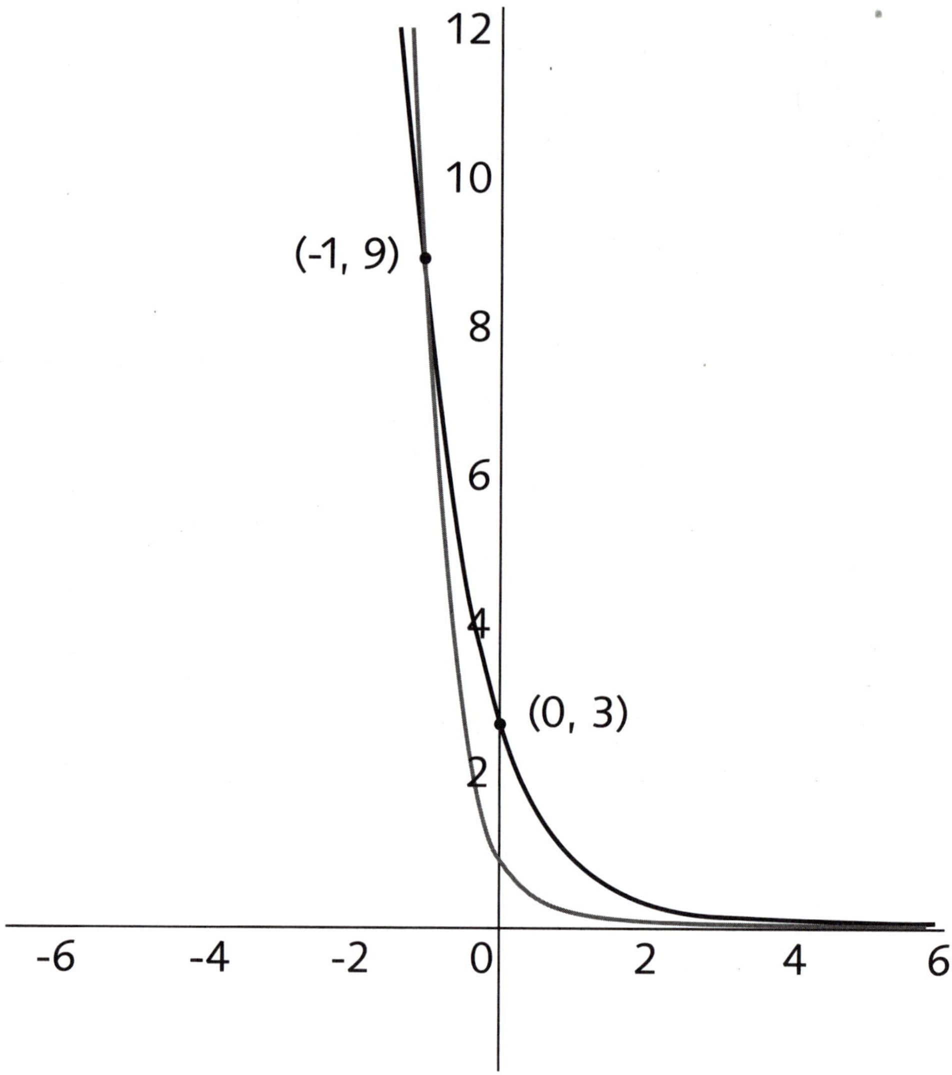

For use with Chapter 6, Question 3, section Exponential and Logarithmic Functions.

Graphs of Two Mystery Functions

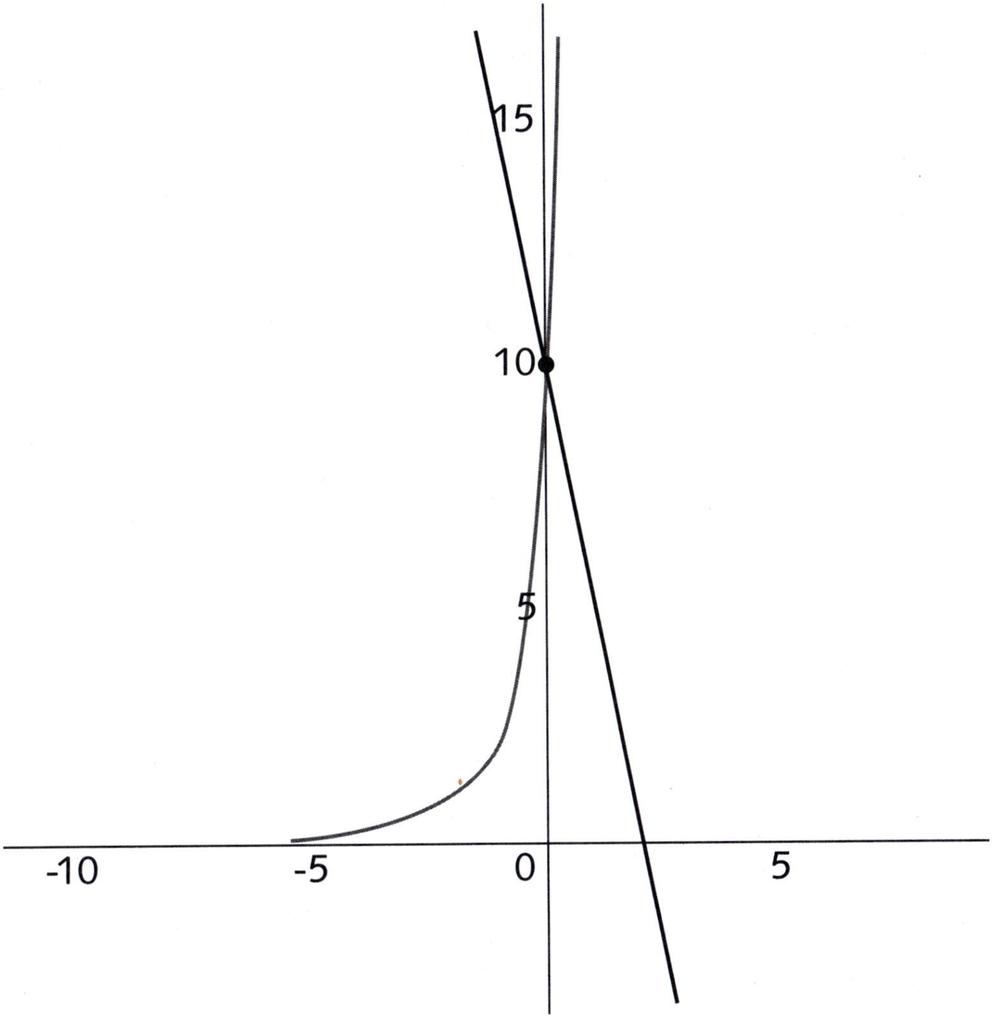

For use with Chapter 6, Question 6, section Exponential and Logarithmic Functions.

Reproducible 22

Graphs of Three Mystery Functions

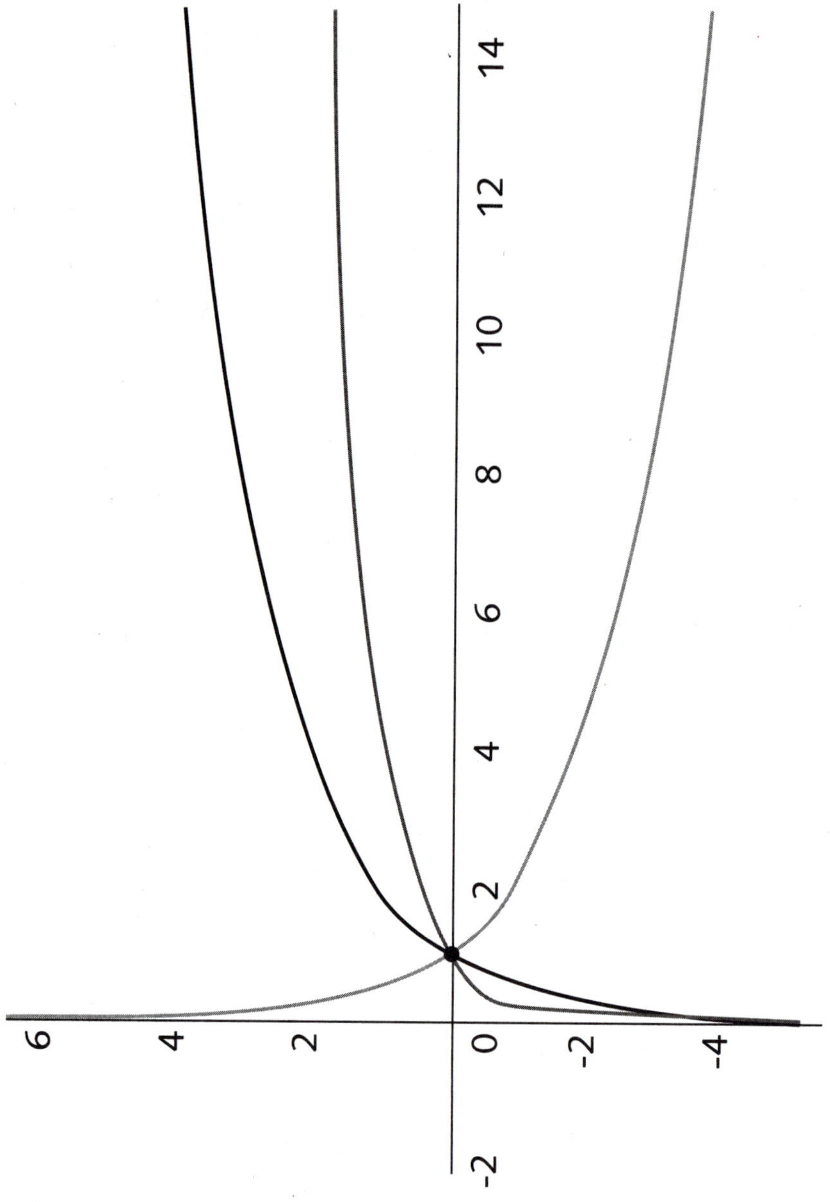

For use with Chapter 6, Question 11, section Exponential and Logarithmic Functions.

Reproducible 23

$y = \log_B(ab^x)$ Mystery Graphs

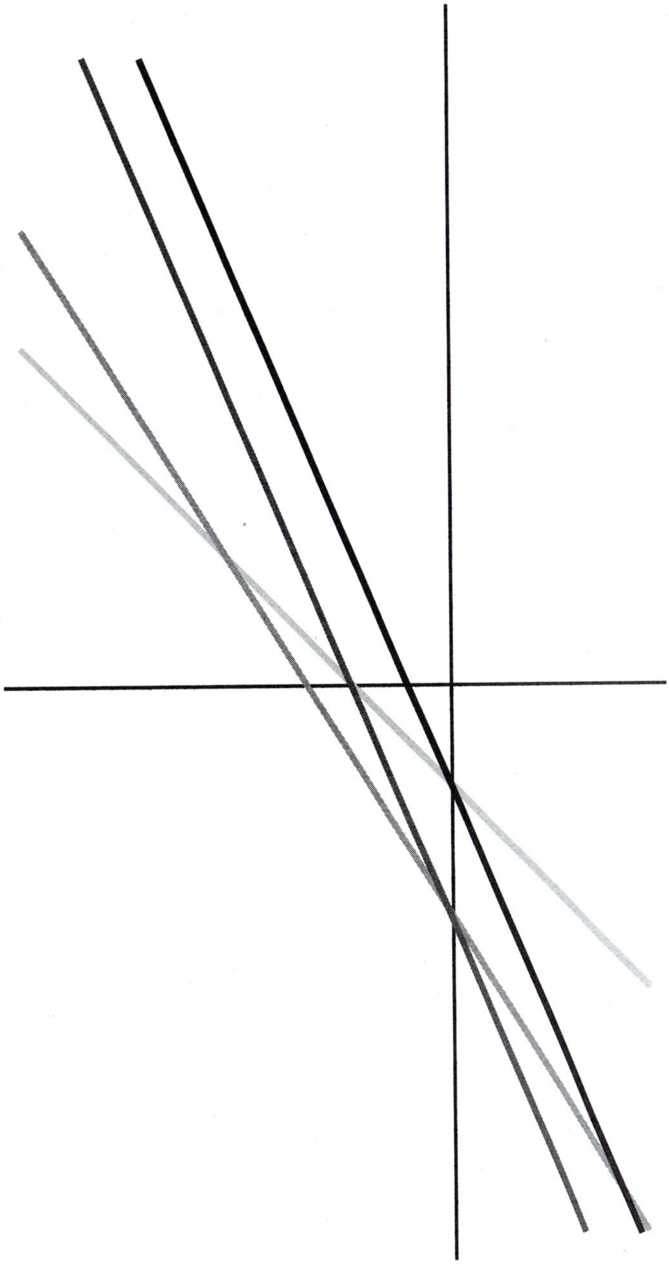

For use with Chapter 6, Question 15, section Exponential and Logarithmic Functions.

Reproducible 24

Mystery Graphs of $y = f(x)$ and $y = 2f(x)$

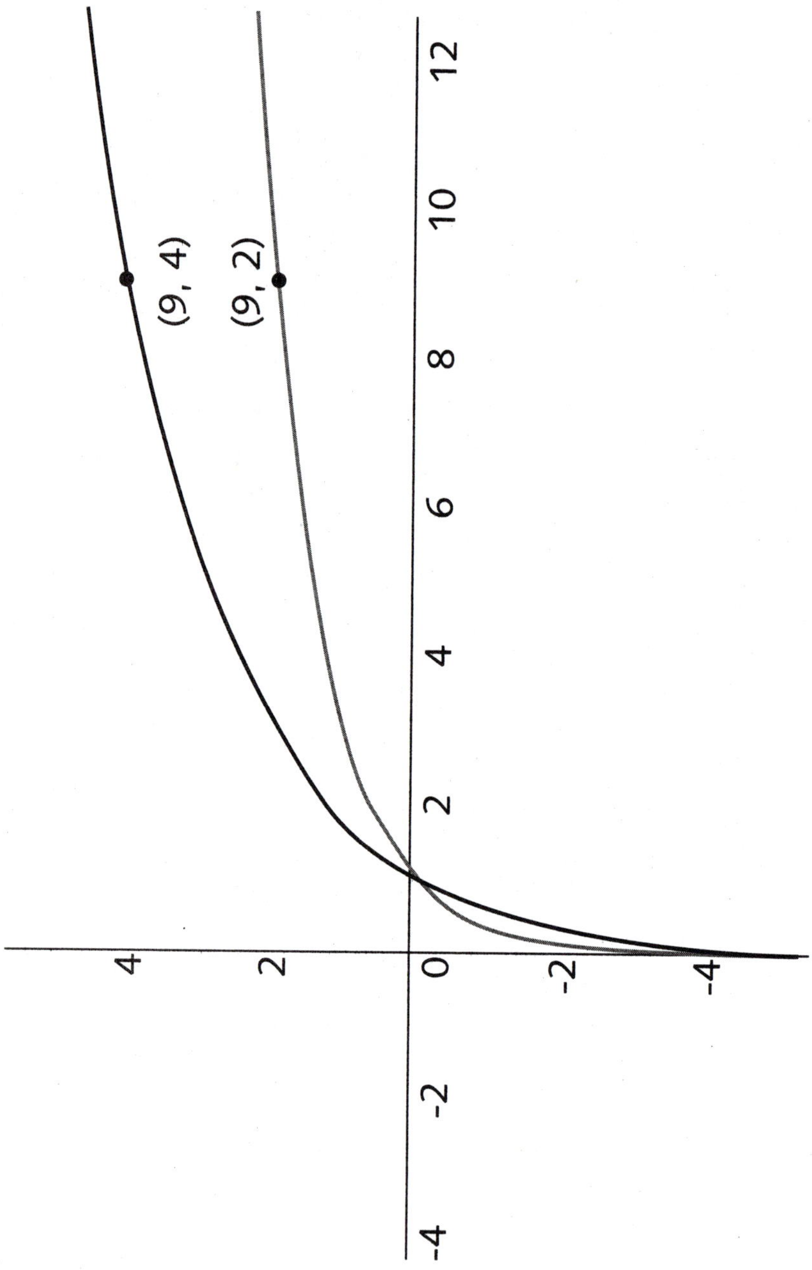

For use with Chapter 6, Question 16, section Exponential and Logarithmic Functions.

Reproducible 25

Representations of Polynomial $P(x)$ and $Q(x)$

$P(x) = 2x^2 - 13x + 18$ $Q(x) =$

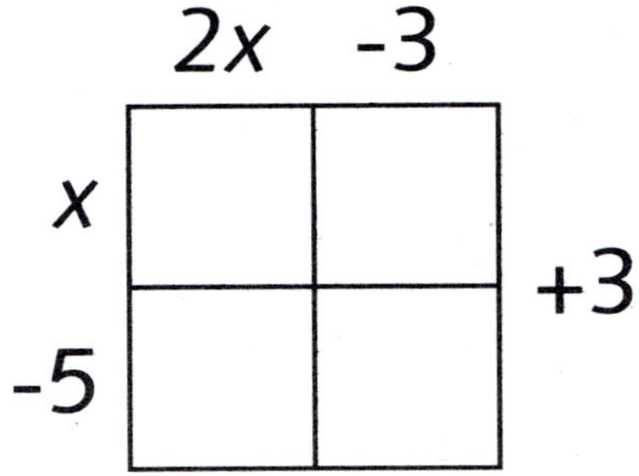

For use with Chapter 6, Question 3, section Polynomial Functions.

Reproducible 26

Relationships between a Cubic and Linear Function

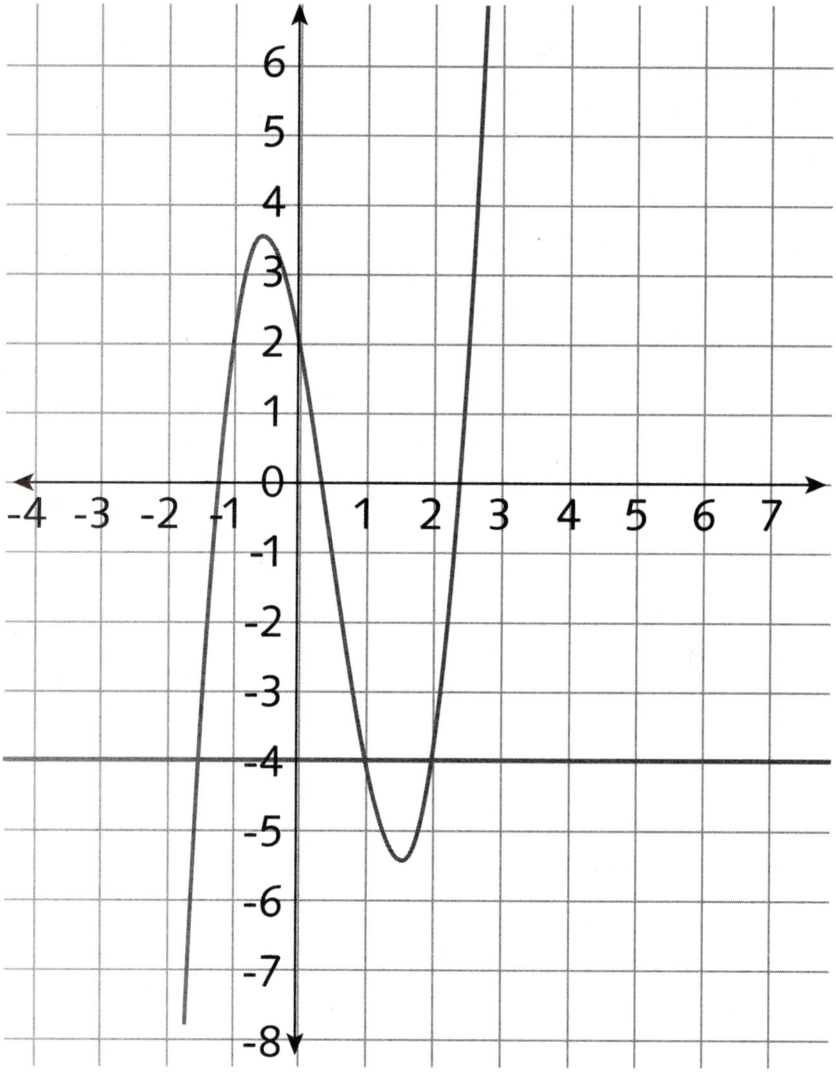

For use with Chapter 6, Questions 6 and 7, section Polynomial Functions.

Reproducible 27

Polynomial $P(x)$ Area Model

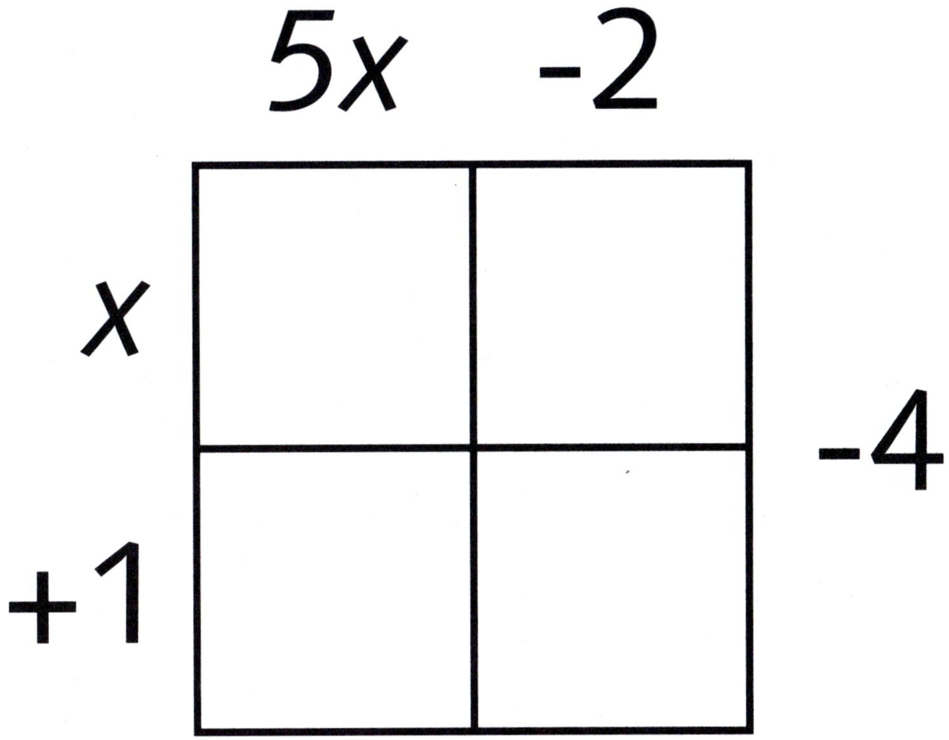

For use with Chapter 6, Question 10, section Polynomial Functions.

Reproducible 28

Selina's Graph and Table

$P(x)$
$(x-1)(x+2)-1$
$(x-0)(x+1)-3$
$(x-2)(x+3)+3$
$(\ ?\ )(\ ?\ )-?$
$(\ ?\ )(\ ?\ )-?$

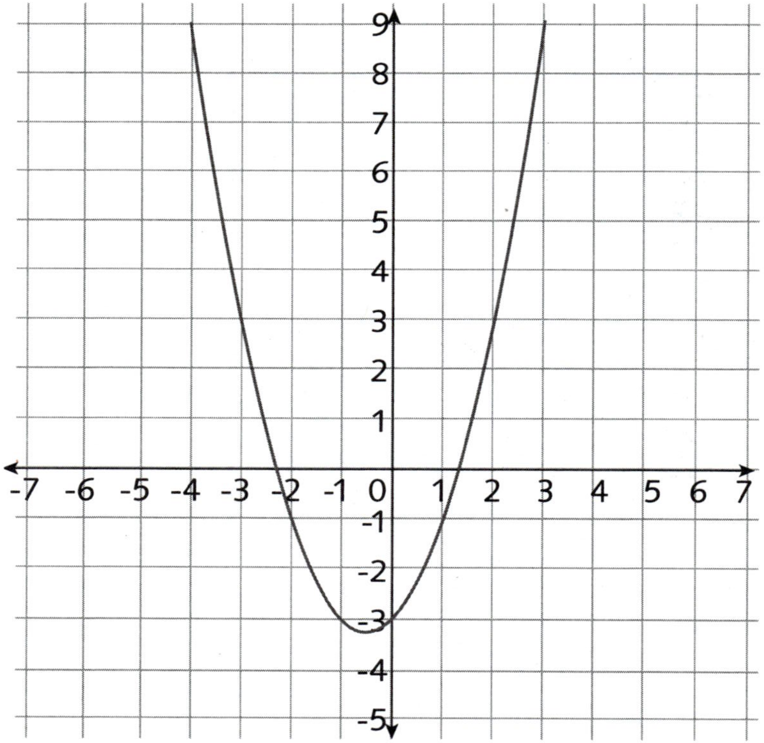

For use with Chapter 6, Question 12, section Polynomial Functions.

Reproducible 29

Graph of $g(x) = 2x^3 + 9x^2 + 3x - 4$

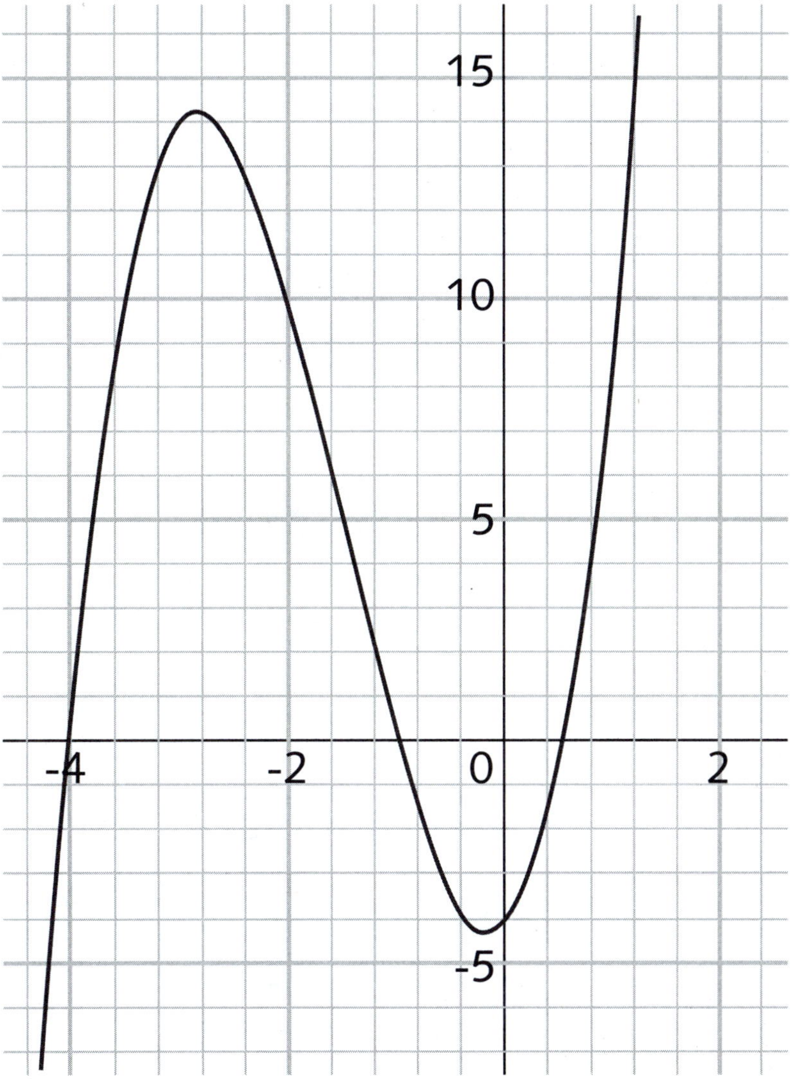

For use with Chapter 6, Question 14, section Polynomial Functions.

REPRODUCIBLE 30

Unit Circles

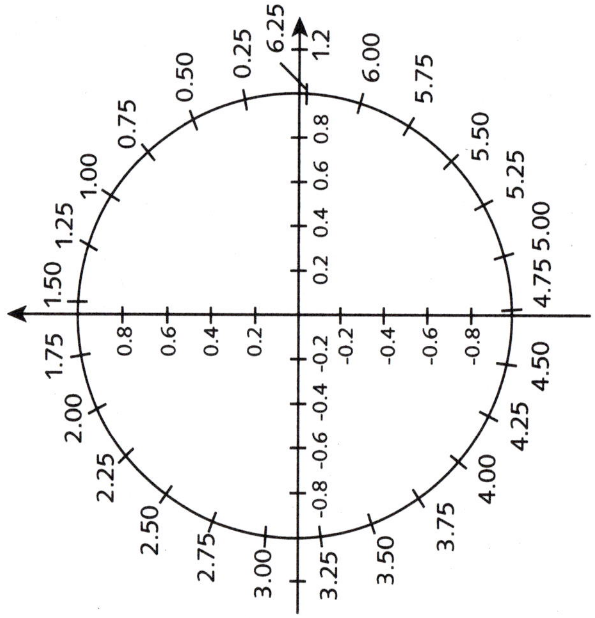

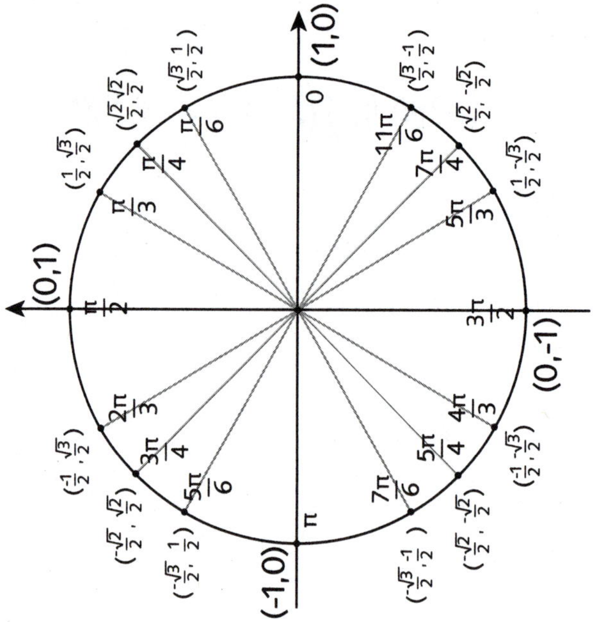

For use with Chapter 6, Question 7, section Trigonometric Functions.

REPRODUCIBLE 31

Abe's and Bea's Calculator Keystrokes

Set I	Set II
[$\tan^{-1}$] [tan] a	[tan] [$\tan^{-1}$] a
[$\sin^{-1}$] [sin] a	[sin] [$\sin^{-1}$] a
[$\cos^{-1}$] [cos] a	[cos] [$\cos^{-1}$] a

For use with Chapter 6, Question 10, section Trigonometric Functions.

REPRODUCIBLE 32

Special Right Triangle Relationships

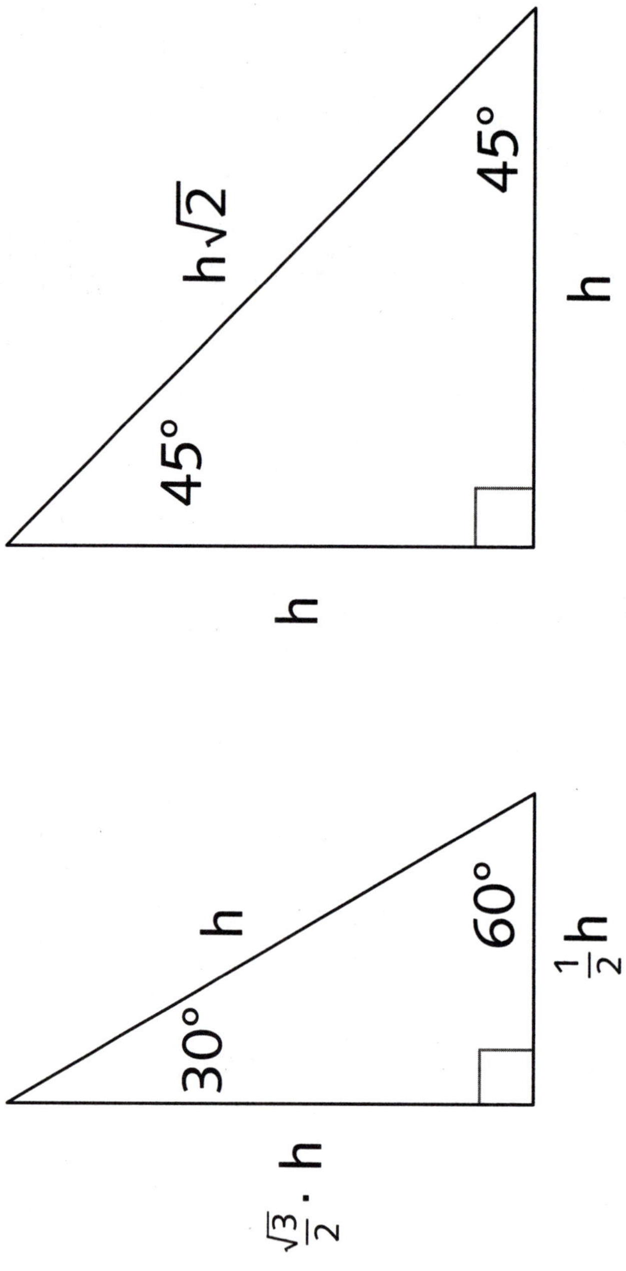

For use with Chapter 6, Question 14, section Trigonometric Functions.

Reproducible 33

Sinusoidal Graph

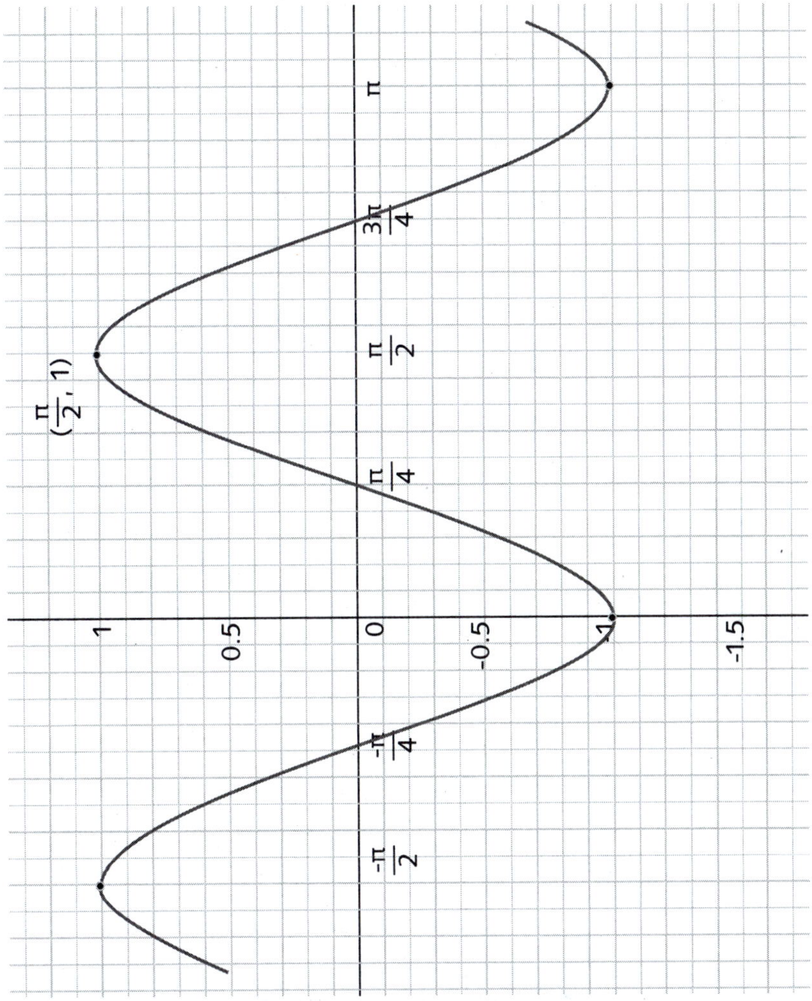

For use with Chapter 6, Question 15, section Trigonometric Functions.

REPRODUCIBLE 34

Composing Functions

x	f(x)
-3	26
-2	17
-1	10
0	5
1	2
2	1
3	2
4	5
5	10

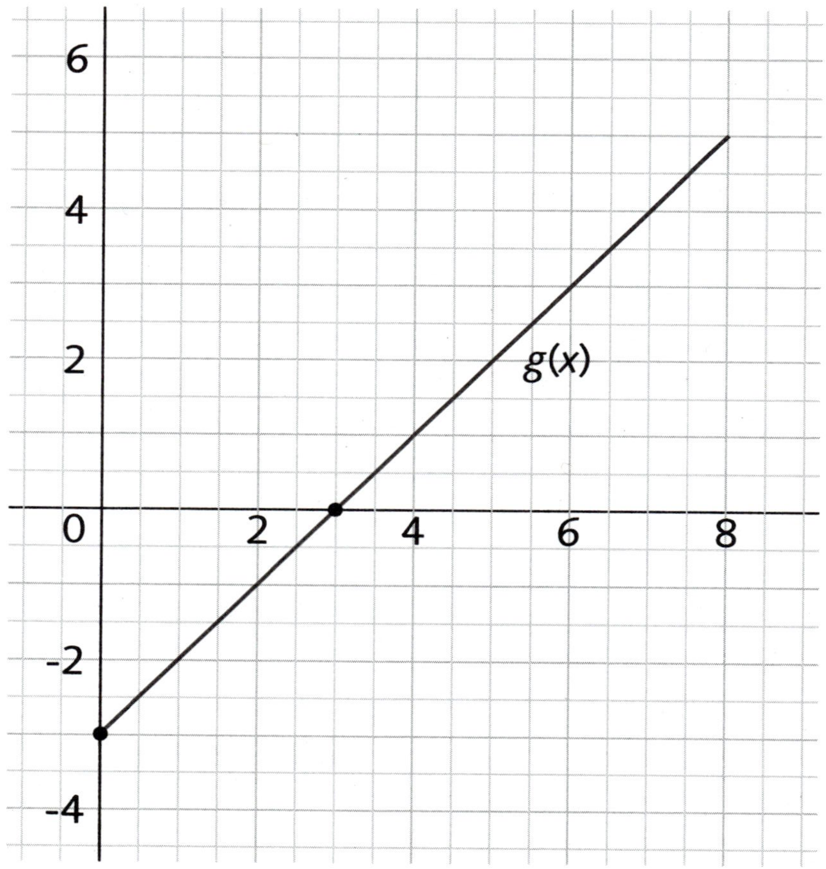

For use with Chapter 6, Question 5, section Inverse Functions and Composites.

Reproducible 35

Marisol's Pen

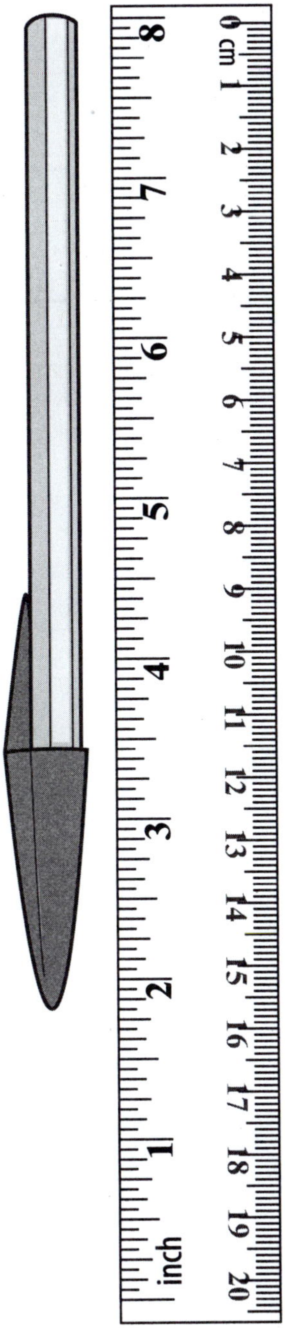

For use with Chapter 7, Question 2, section Measurement.

REPRODUCIBLE 36

Geometric Shapes A, B, C, and D

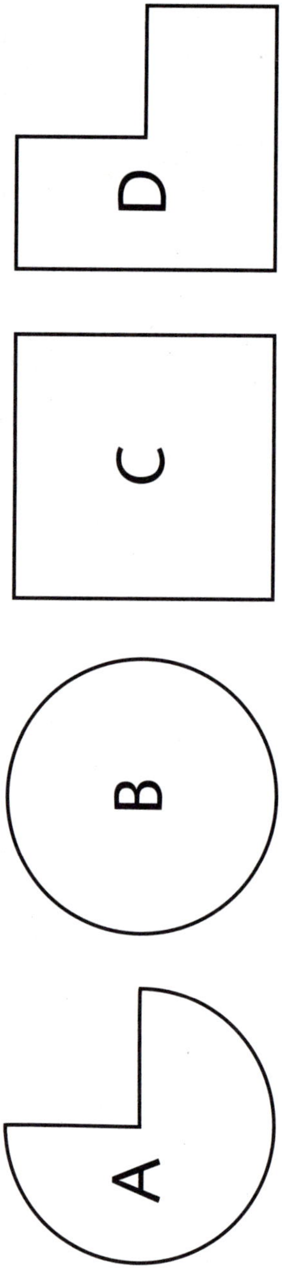

For use with Chapter 7, Question 3, section Measurement.

Reproducible 37

Four Triangles

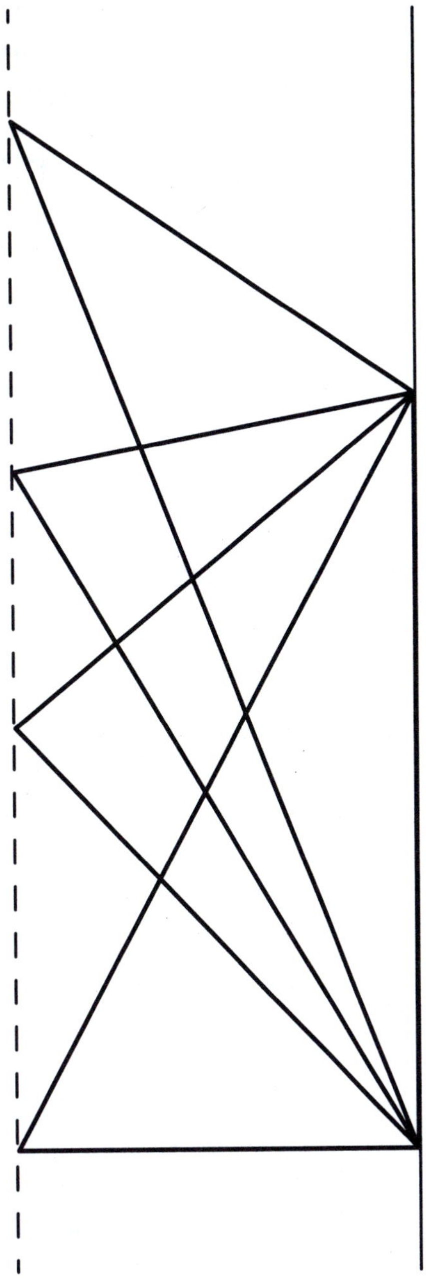

For use with Chapter 7, Question 5, section Measurement.

REPRODUCIBLE 38

Sector of a Circle

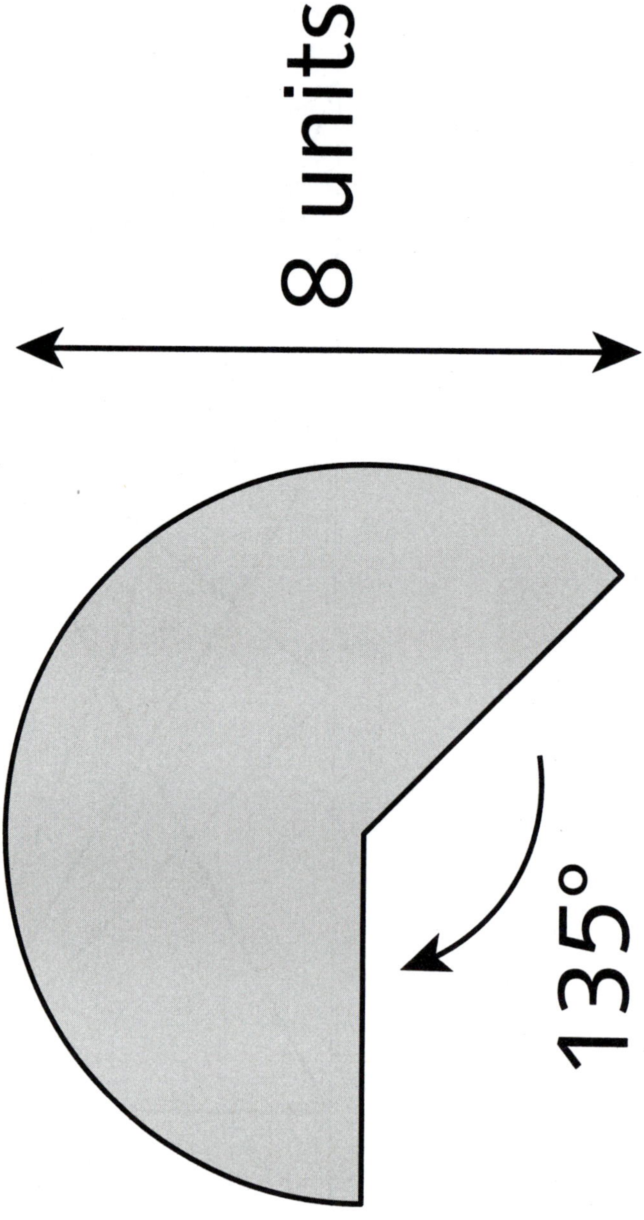

For use with Chapter 7, Question 6, section Measurement.

Reproducible 39

Figures A, B, C, D, and E

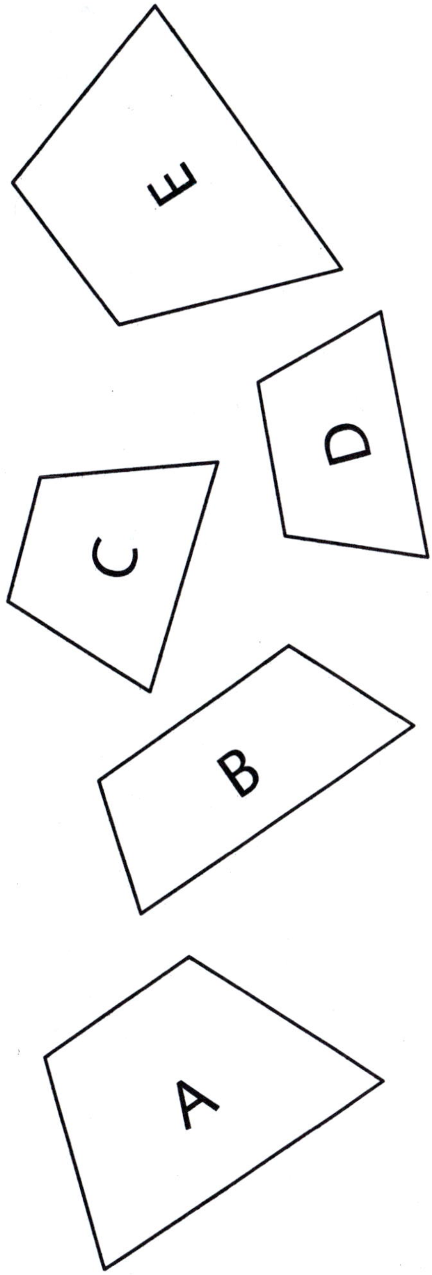

For use with Chapter 7, Question 4, section Congruence and Similarity.

Reproducible 40

Overlapping Triangles

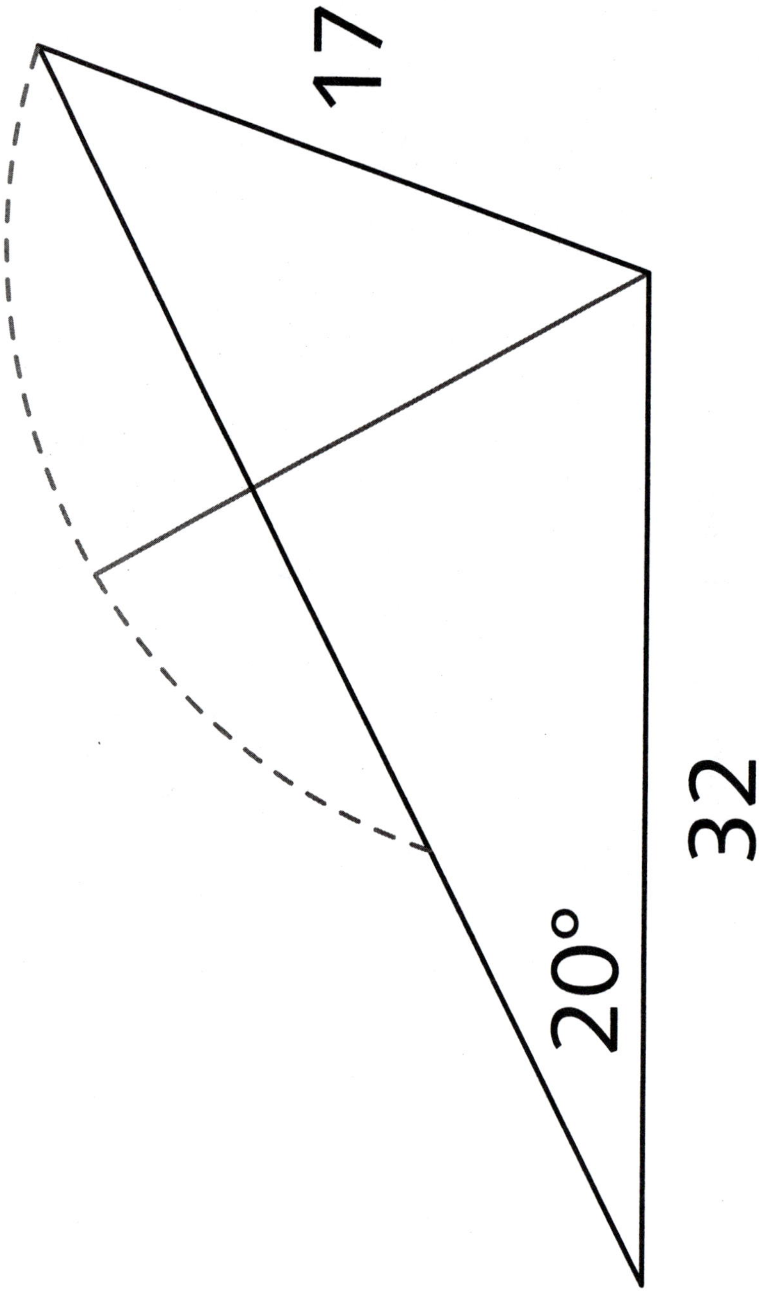

For use with Chapter 7, Question 6, section Congruence and Similarity.

Reproducible 41

ΔABC and ΔDCB

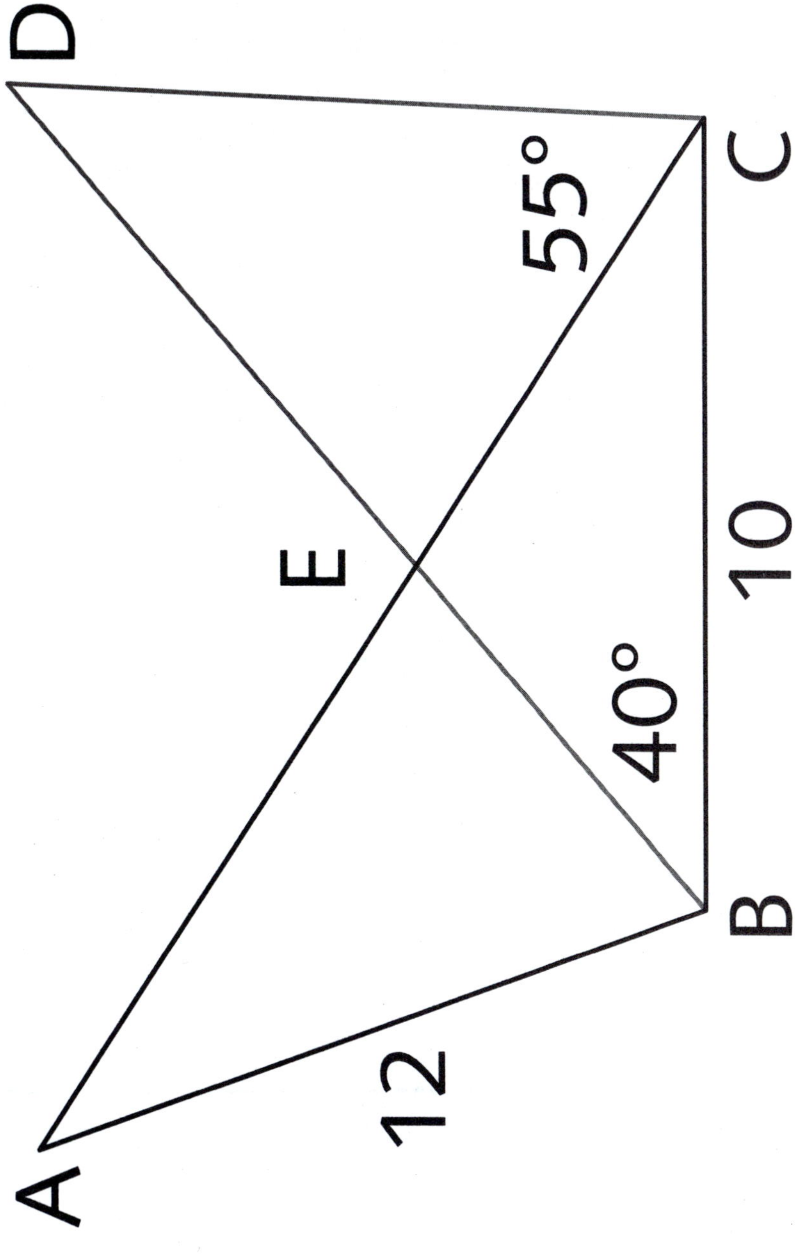

For use with Chapter 7, Question 7, section Congruence and Similarity.

Reproducible 42

Three Right Triangles

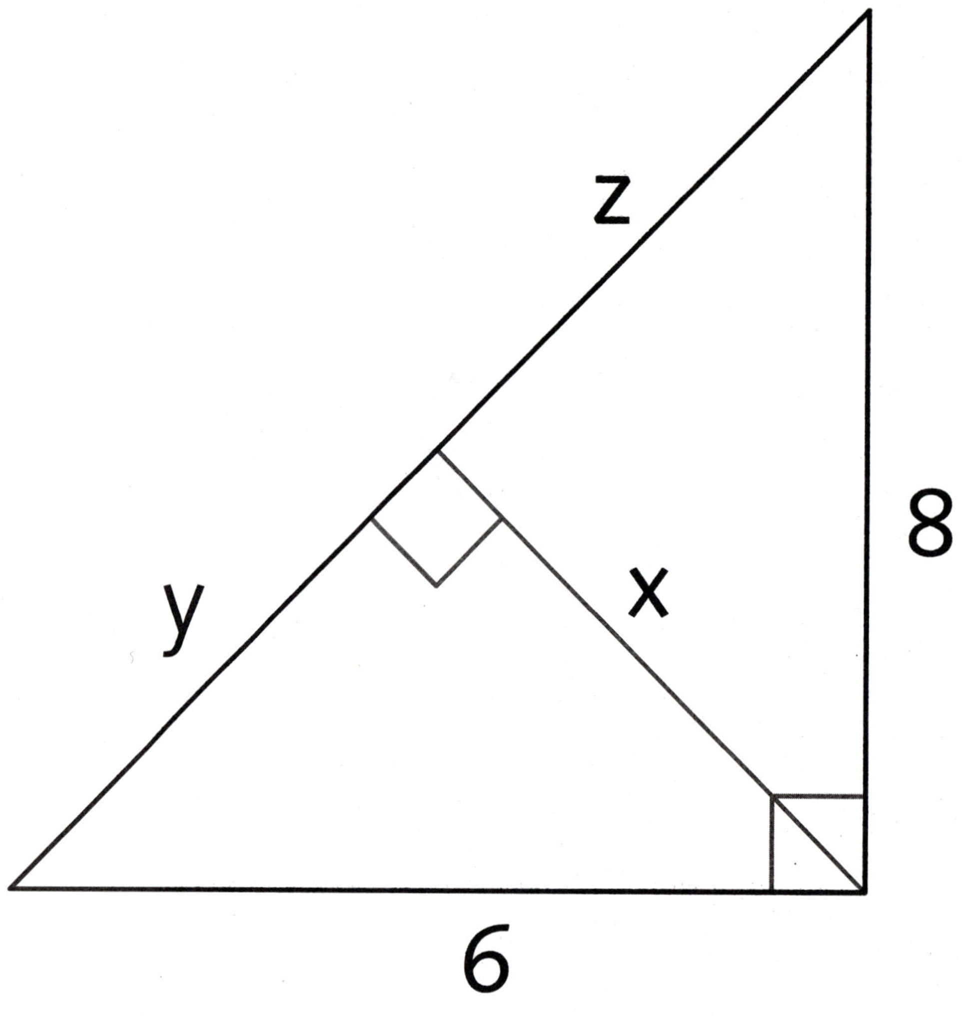

For use with Chapter 7, Question 9, section Congruence and Similarity.

Reproducible 43

Two Versions of 30-60-90 Triangles

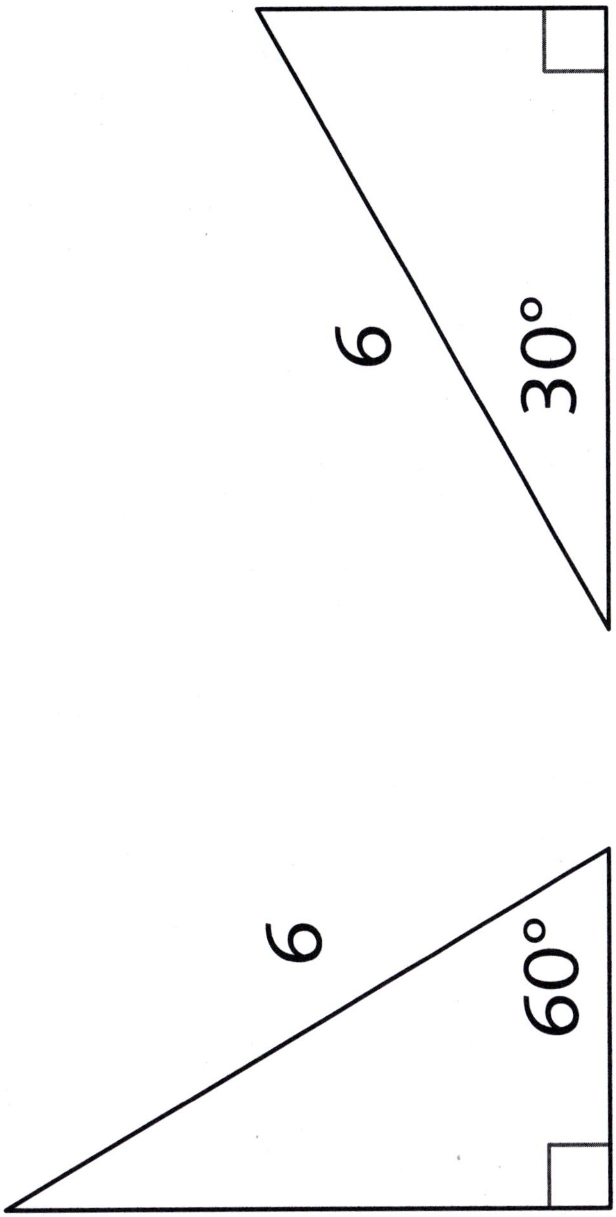

For use with Chapter 7, Question 10, section Congruence and Similarity.

Reproducible 44

$\triangle ABC$ and $\triangle DFE$

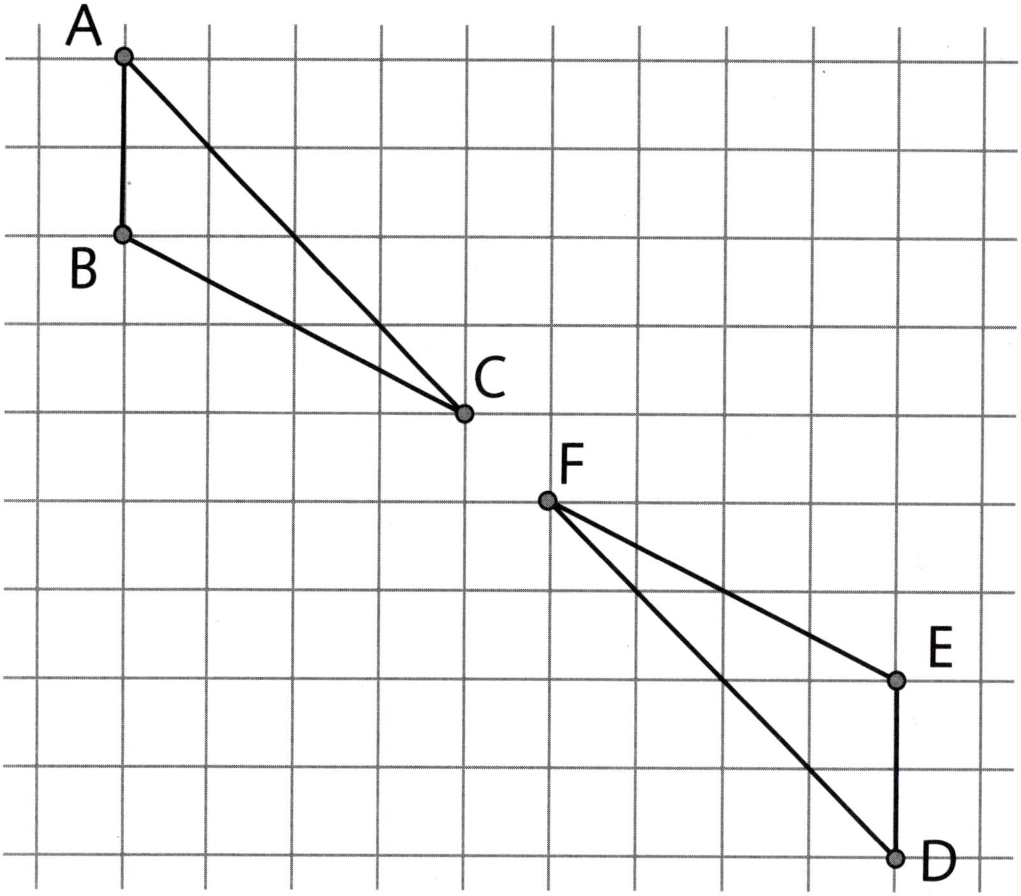

For use with Chapter 7, Question 2, section Transformations.

Reproducible 45

Rotating Quadrilateral

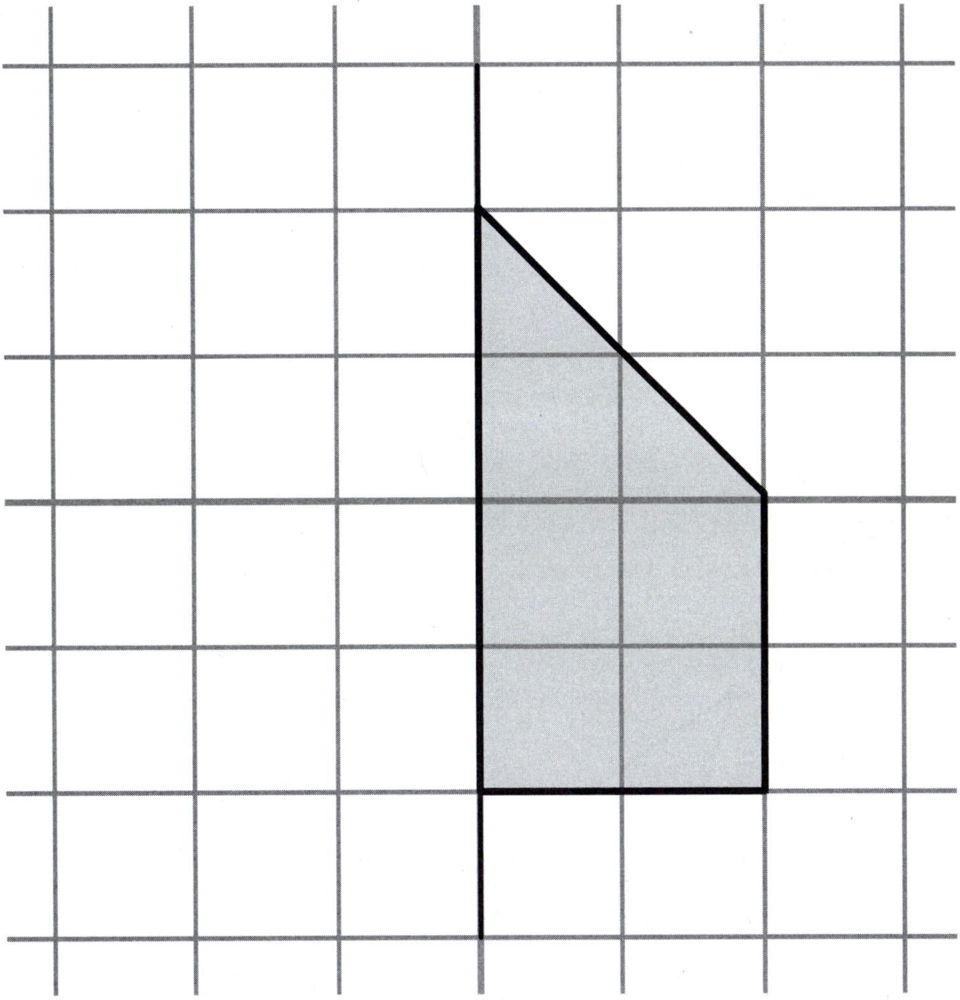

For use with Chapter 7, Question 3, section Transformations.

Reproducible 46

△*ABC* and Point *D*

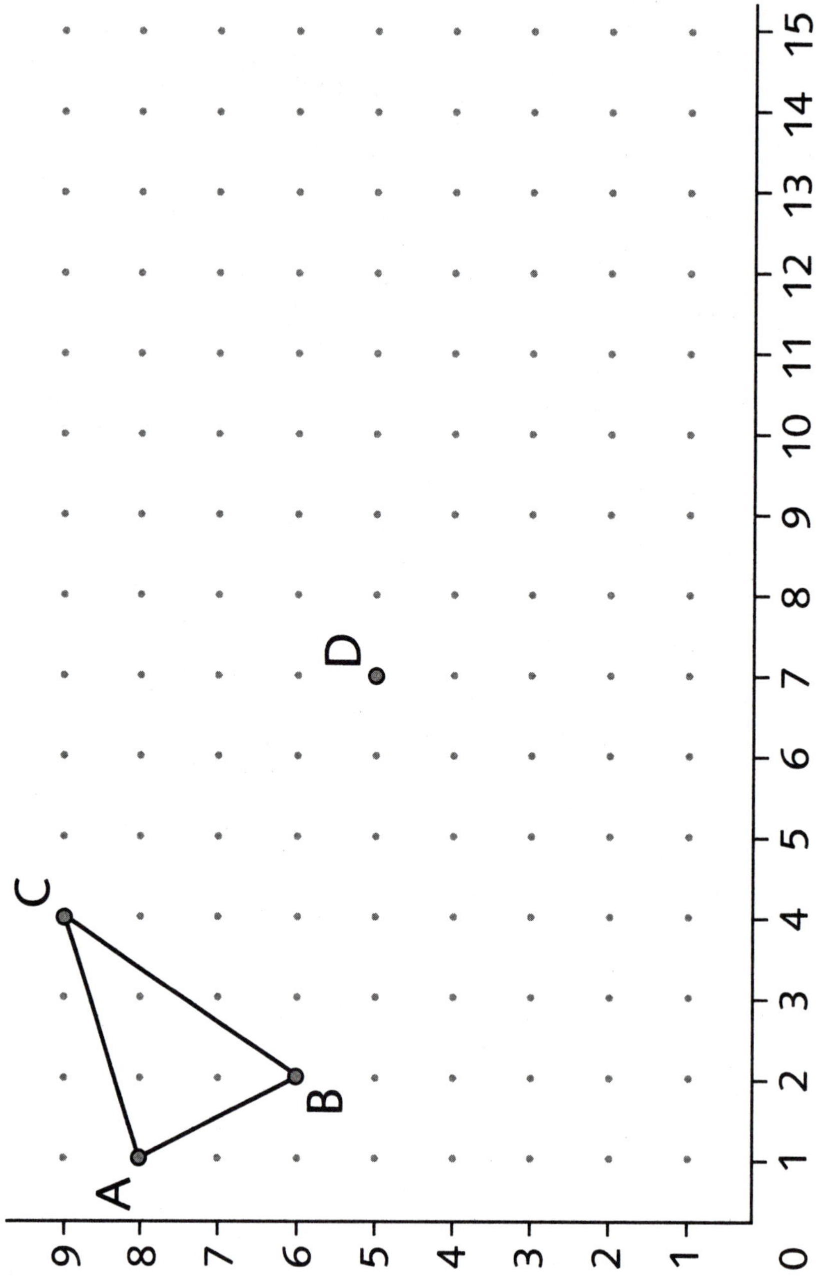

For use with Chapter 7, Question 13, section Transformations.

Reproducible 47

△ABC

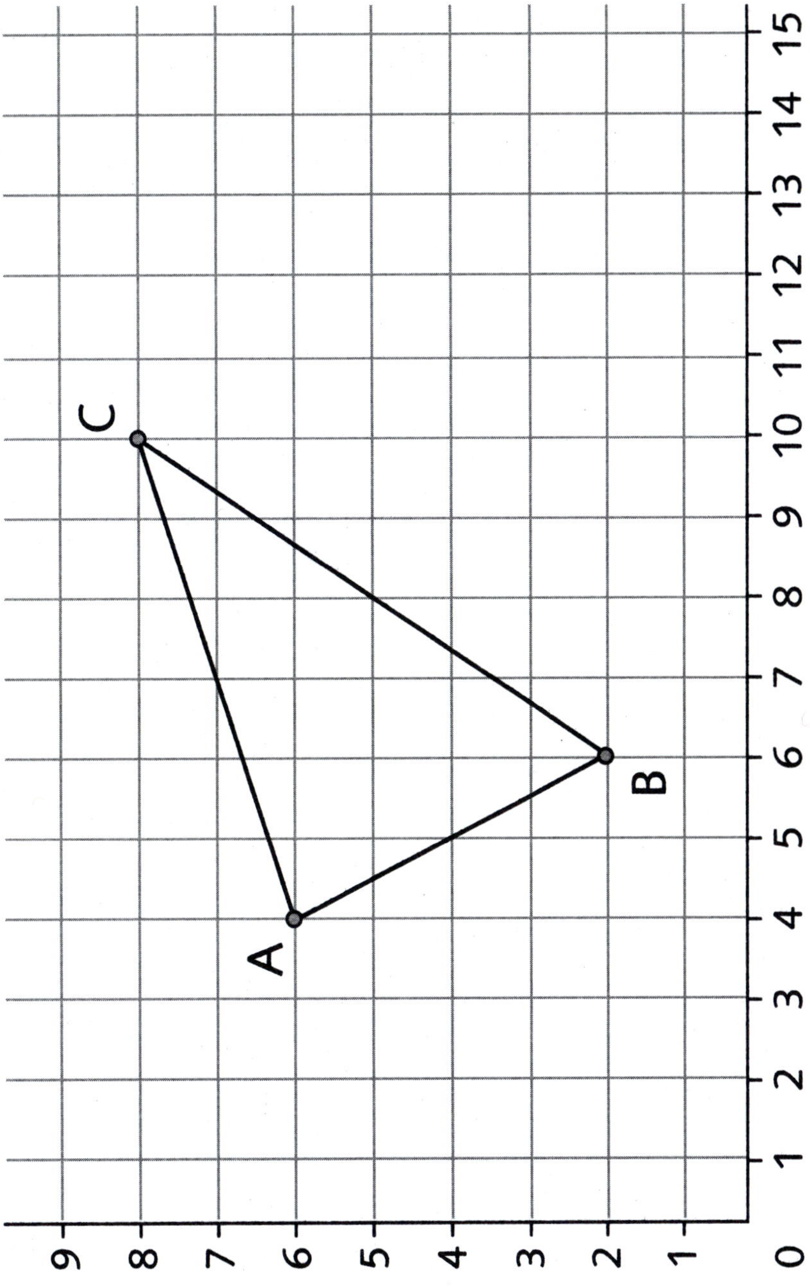

For use with Chapter 7, Question 14, section Transformations.

Square *ABCD* and $\triangle EGF$

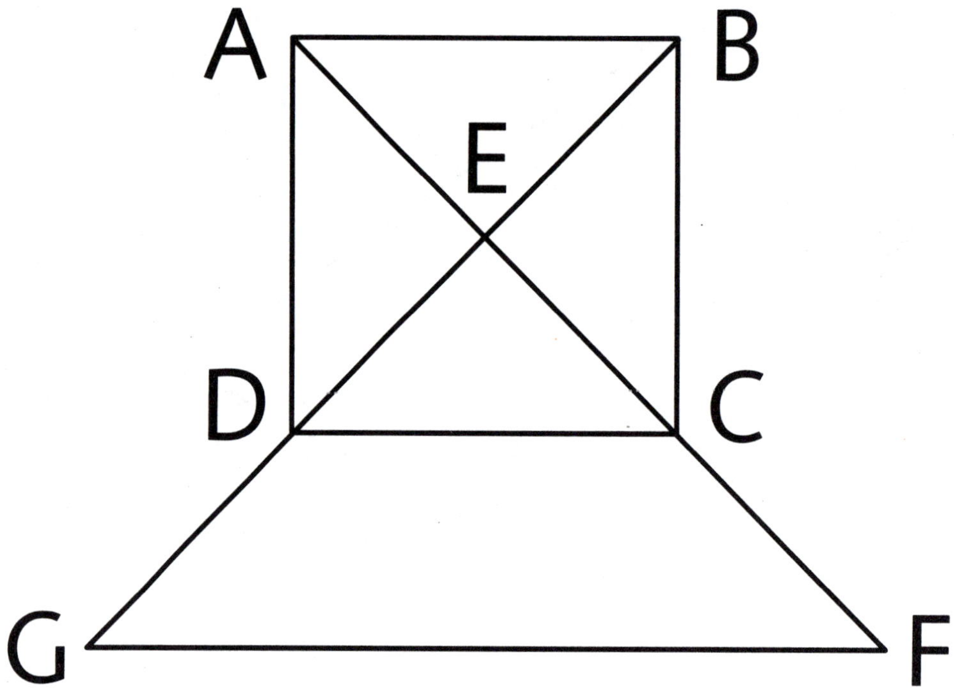

For use with Chapter 7, Question 8, section Polygons.

REPRODUCIBLE 49

Linear Angles

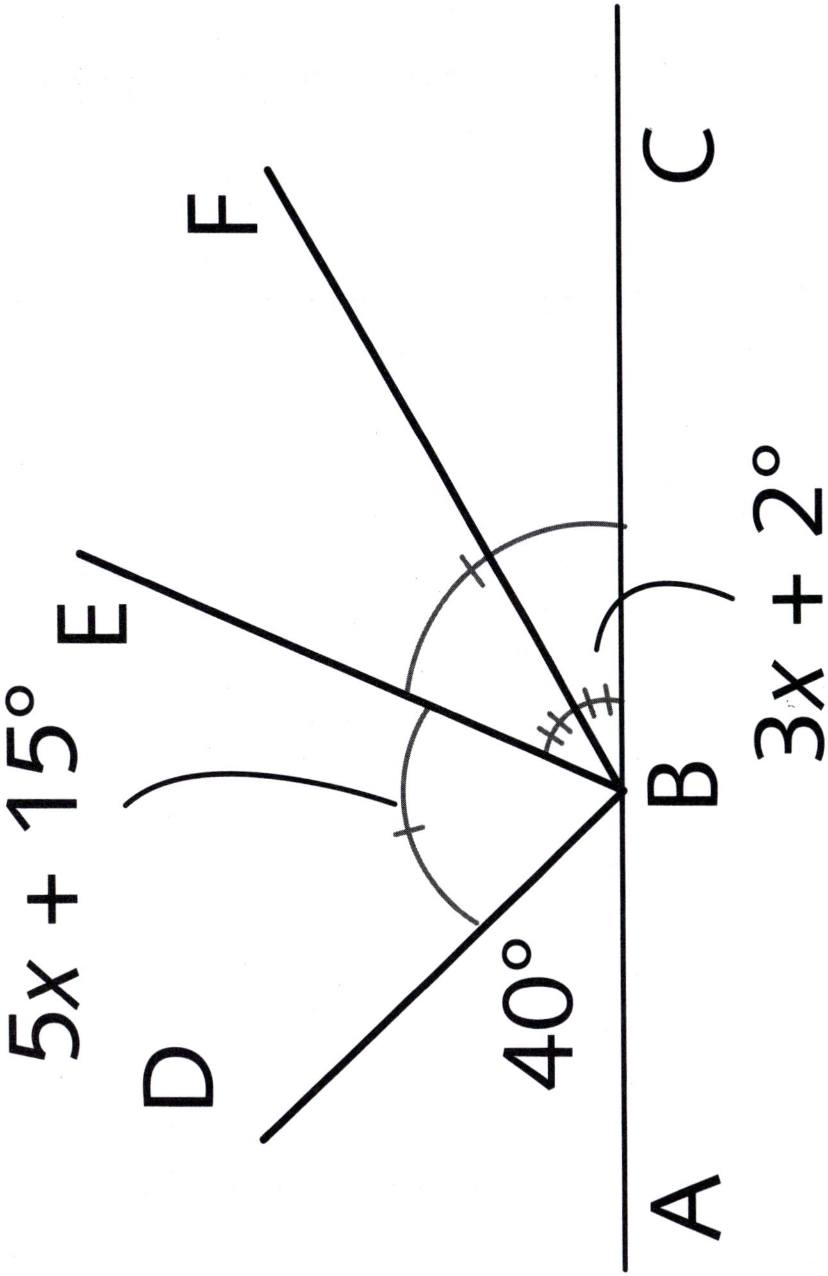

For use with Chapter 7, Question 9, section Polygons.

REPRODUCIBLE 50

Square Cut by Two Perpendicular Lines

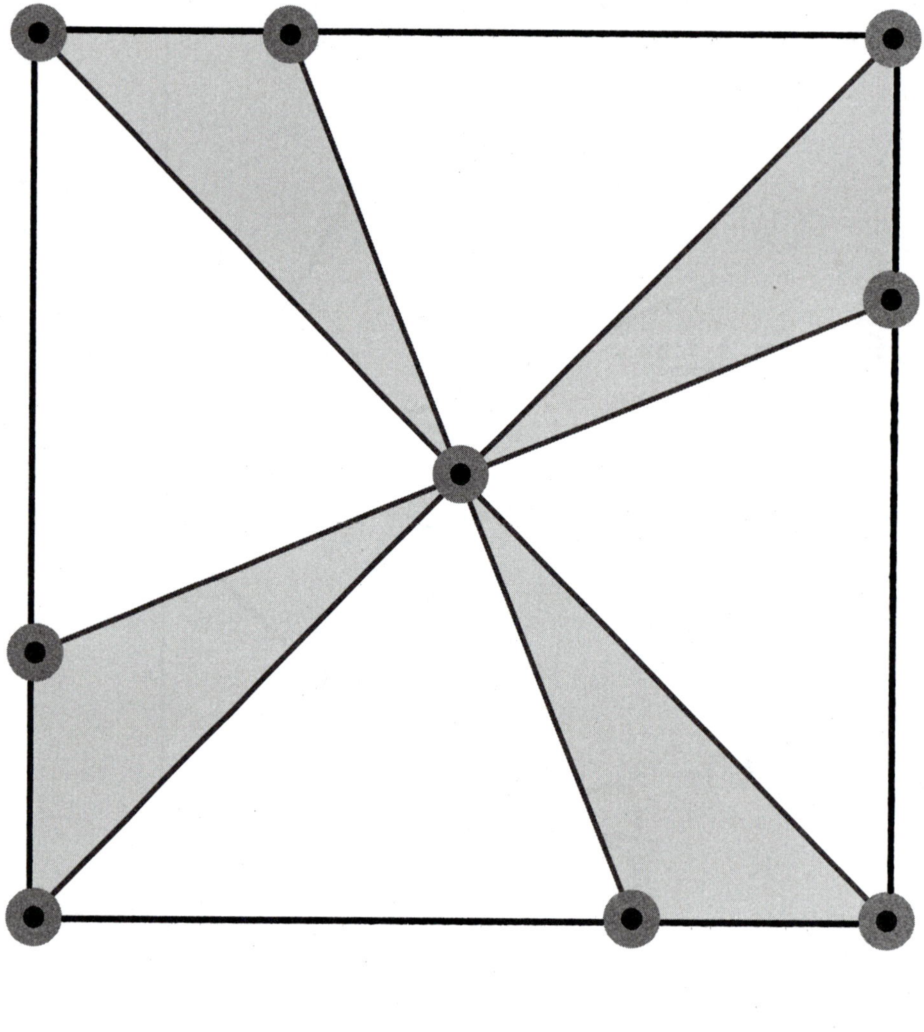

For use with Chapter 7, Question 10, section Polygons.

Reproducible 51

Right Triangle *ADB*

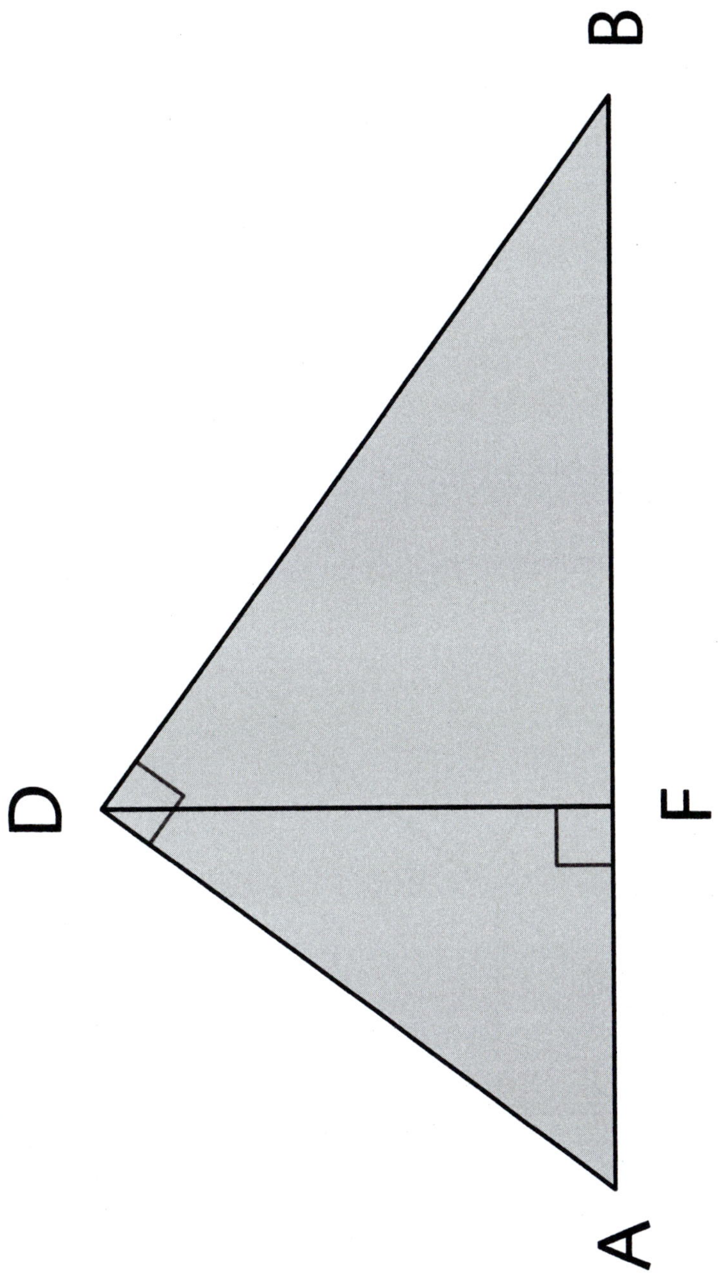

For use with Chapter 7, Question 3, section Right Triangles and Trigonometry.

Similar Right Triangles

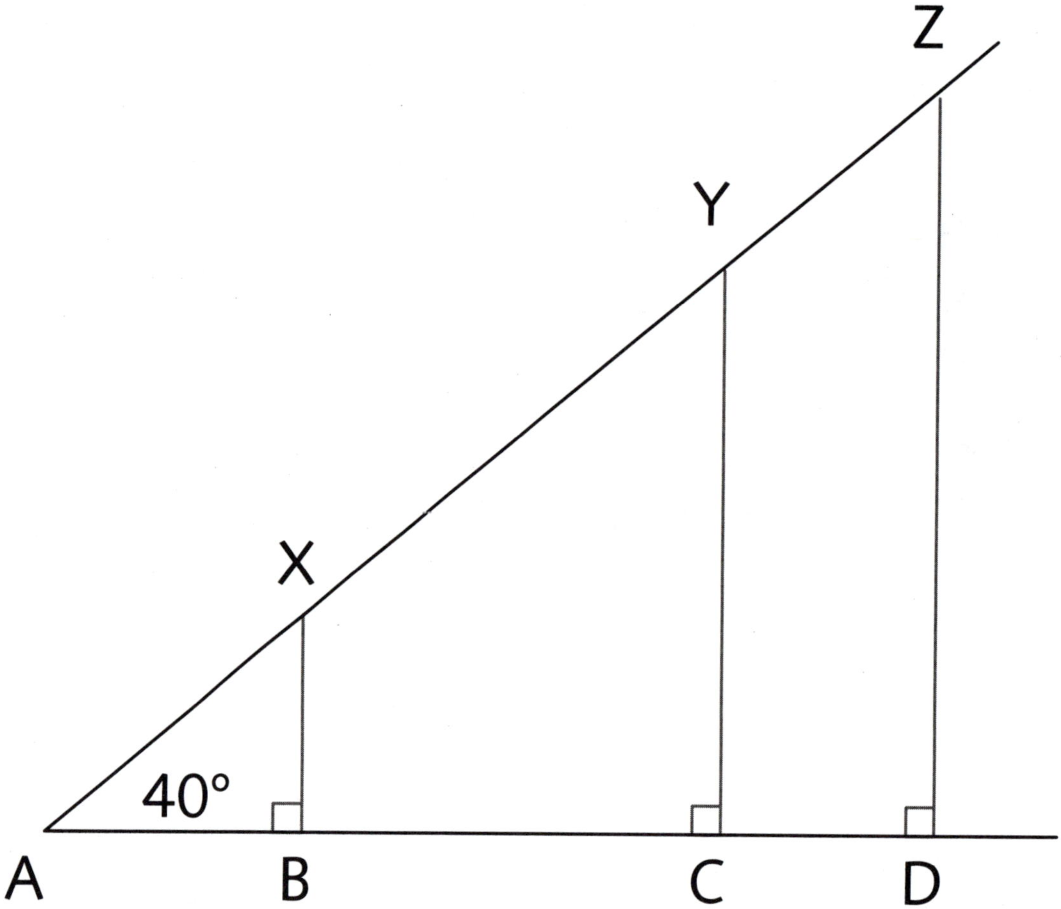

For use with Chapter 7, Question 4, section Right Triangles and Trigonometry.

Reproducible 53

Labeled Right Triangle

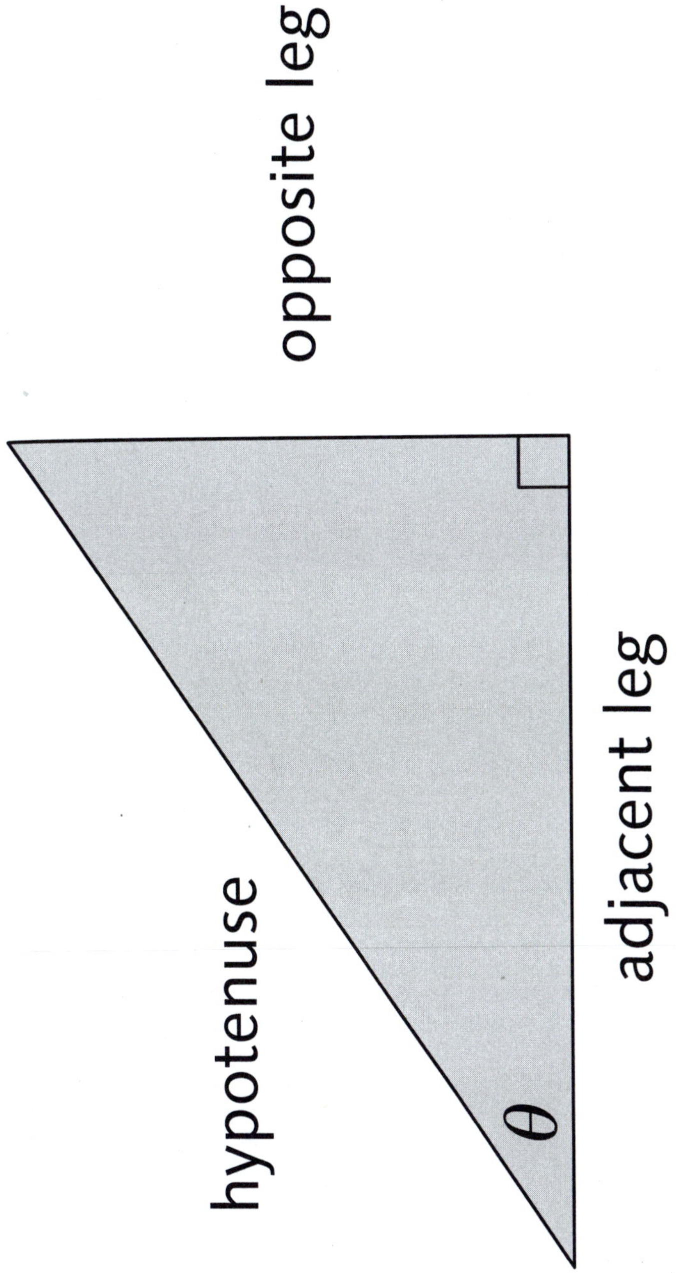

For use with Chapter 7, Question 14, section Right Triangles and Trigonometry.

Reproducible 54

$\triangle ABC$ with Altitude h

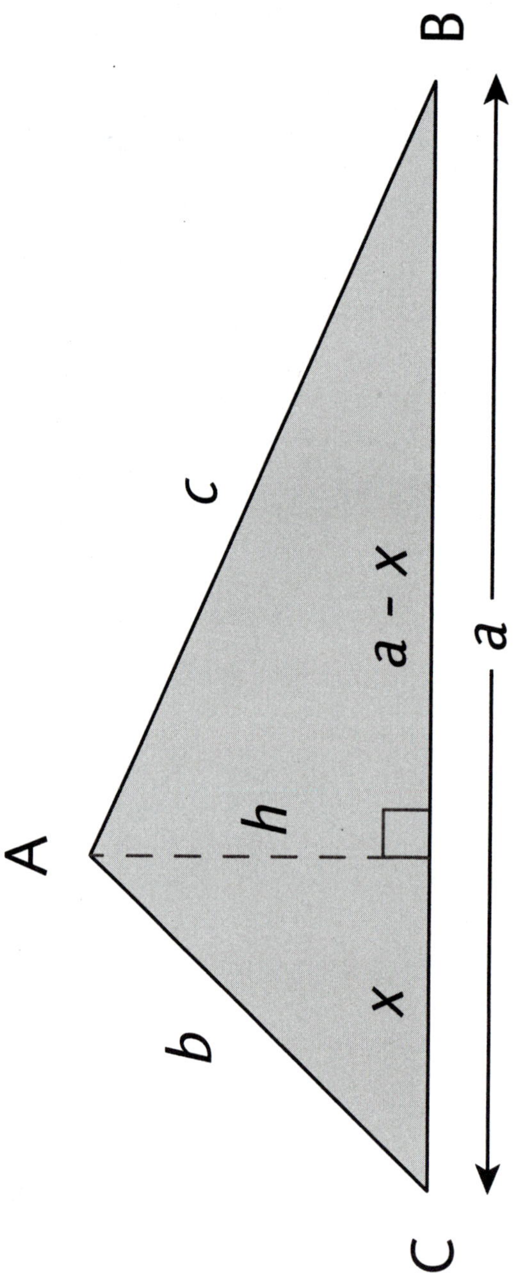

For use with Chapter 7, Question 15, section Right Triangles and Trigonometry.

Reproducible 55

GQ's Triangle

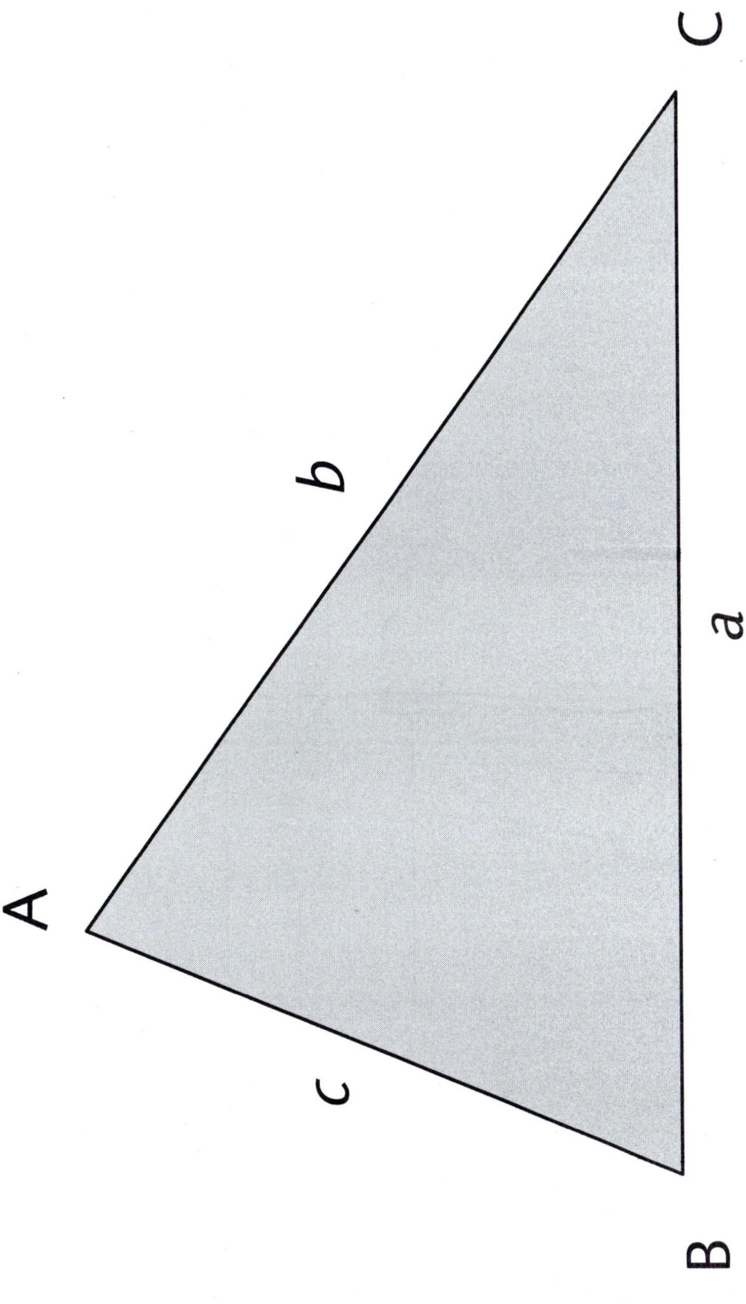

For use with Chapter 7, Question 16, section Right Triangles and Trigonometry.

Reproducible 56

Data Display of Sugar Content of Adult versus Children's Cereal

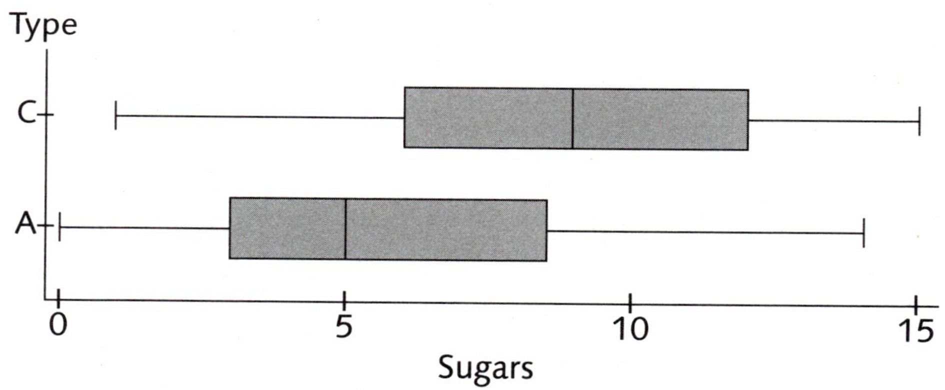

Type	*n*	Mean	Standard Deviation	Min	Q1	Median	Q3	Max
A	44	5.6363636	4.2266646	0	3	5	8.5	14
C	33	8.6666667	4.1281554	1	6	9	12	15

For use with Chapter 8, Question 4, section Interpreting Categorical and Quantitative Data.

REPRODUCIBLE 57

Plot of Sugar in Cereal by Display Shelf

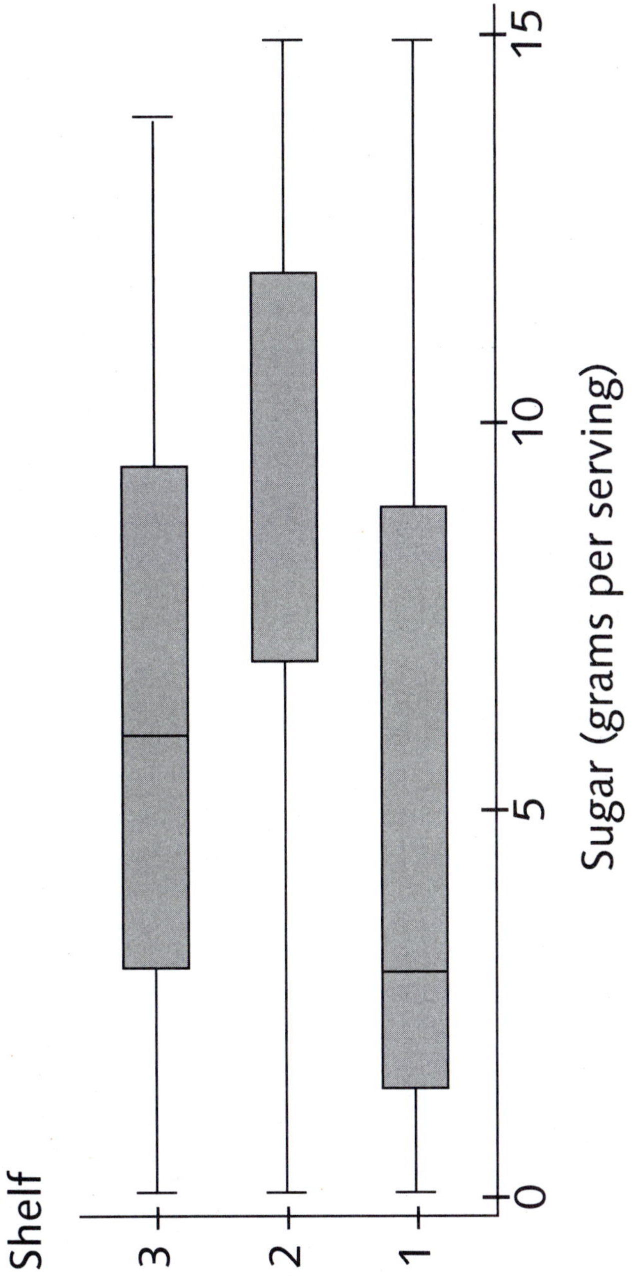

For use with Chapter 8, Question 5, section Interpreting Categorical and Quantitative Data.

REPRODUCIBLE 58

Turtle Weights Dot Plot

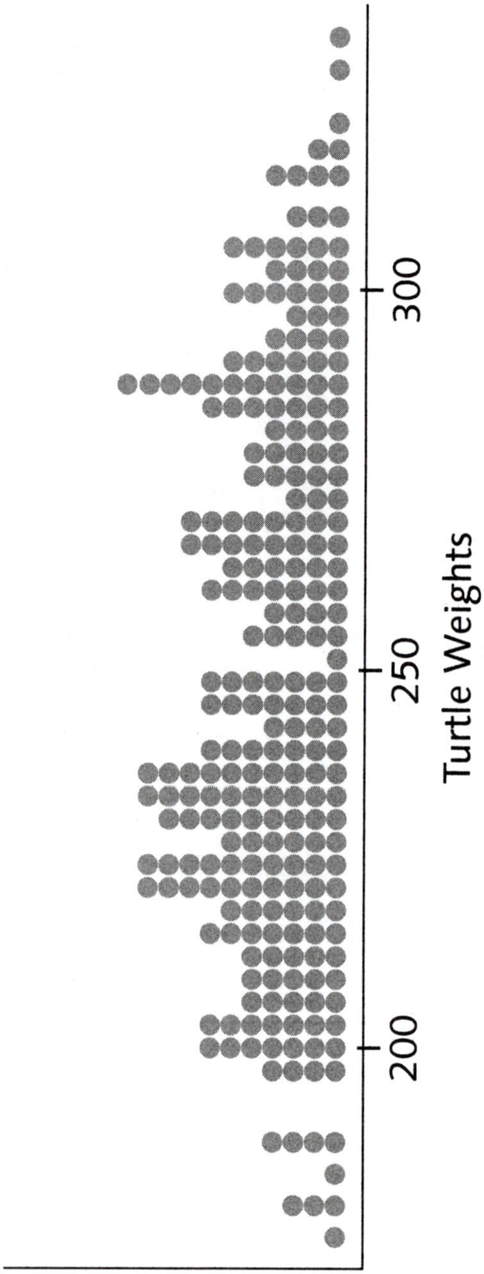

For use with Chapter 8, Question 6, section Interpreting Categorical and Quantitative Data.

REPRODUCIBLE 59

Plot of Baseball Salaries in 2017

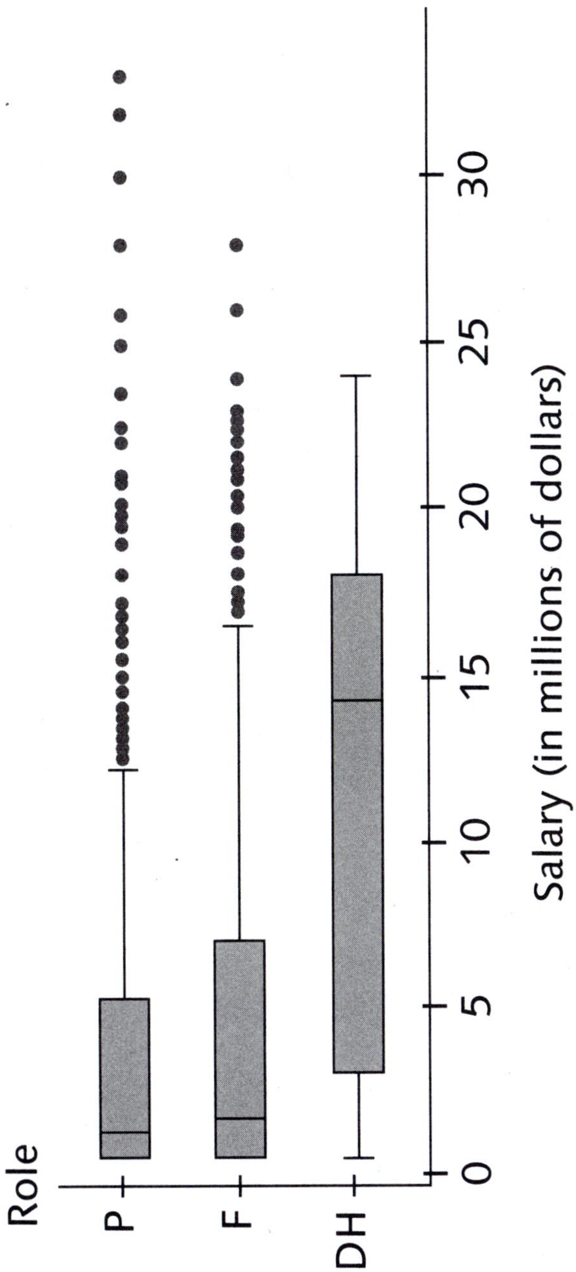

For use with Chapter 8, Question 7, section Interpreting Categorical and Quantitative Data.

Reproducible 60

Data Displays of Cereal Nutrition

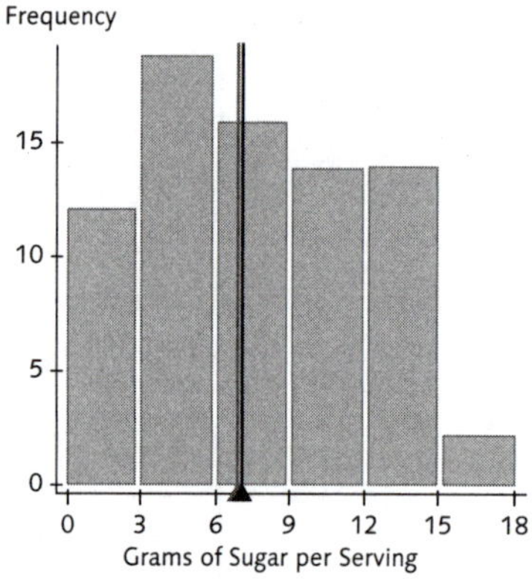

Histogram 1

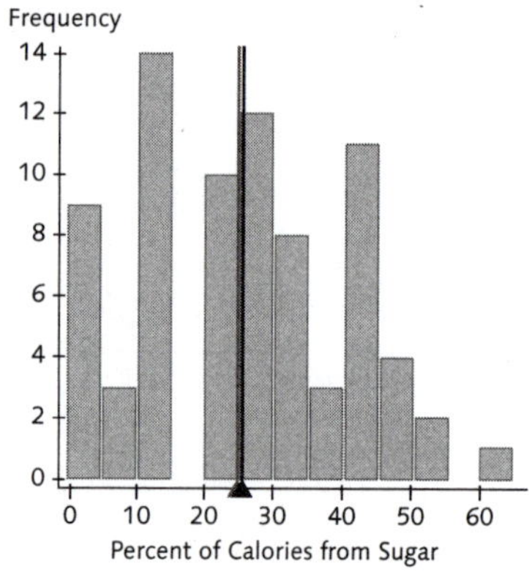

Histogram 2

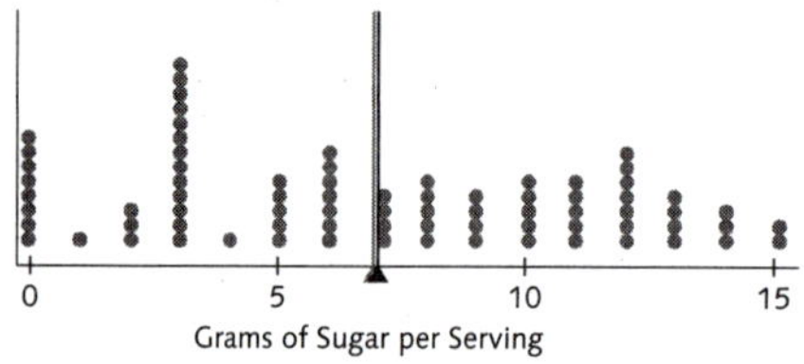

Dot plot 1

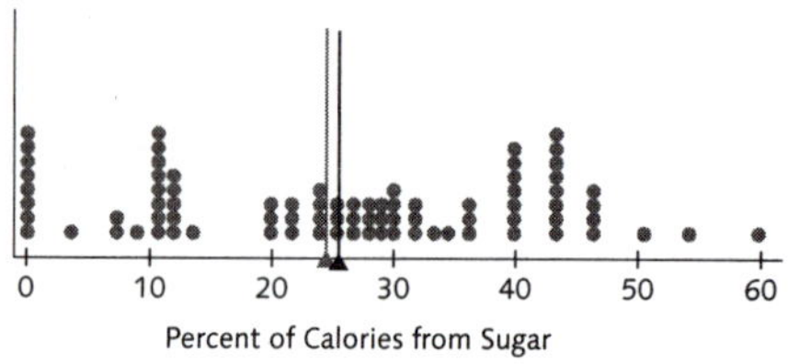

Dot plot 2

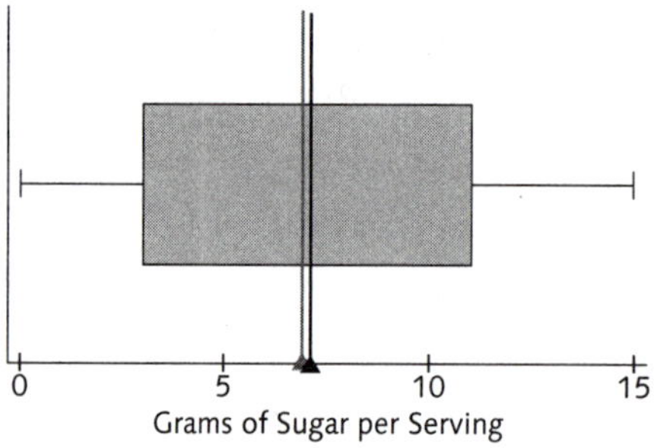

Boxplot 1

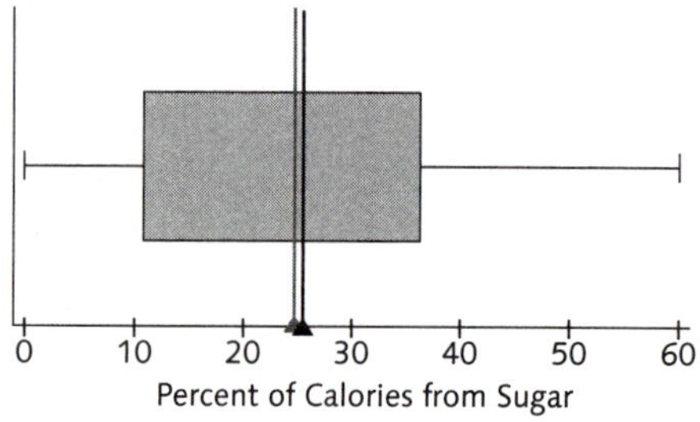

Boxplot 2

Note: The mean and median are plotted on each graph.

For use with Chapter 8, Question 8, section Interpreting Categorical and Quantitative Data.

Reproducible 61

Table of Temperature Statistics by Season

Summary Statistics for Average Temperature (°C)				
Season	Mean (°C)	Standard Deviation (°C)	Median (°C)	IQR (°C)
Fall	3.34	9.29	5.83	15.62
Spring	12.87	6.31	13.85	9.35
Summer	19.60	3.22	20.1	4.80
Winter	-4.32	6.30	-3.67	9.11

For use with Chapter 8, Question 11, section Interpreting Categorical and Quantitative Data.

Sets of Die-Rolling Results

Set I	
Pips	Frequency
1	13
2	7
3	9
4	12
5	13
6	6
Total	60

Set II	
Pips	Frequency
1	130
2	70
3	90
4	120
5	130
6	60
Total	600

Set III	
Pips	Frequency
1	103
2	97
3	99
4	102
5	103
6	96
Total	600

For use with Chapter 8, Question 23, section Interpreting Categorical and Quantitative Data.

Reproducible 63

Nine Point Scatterplot

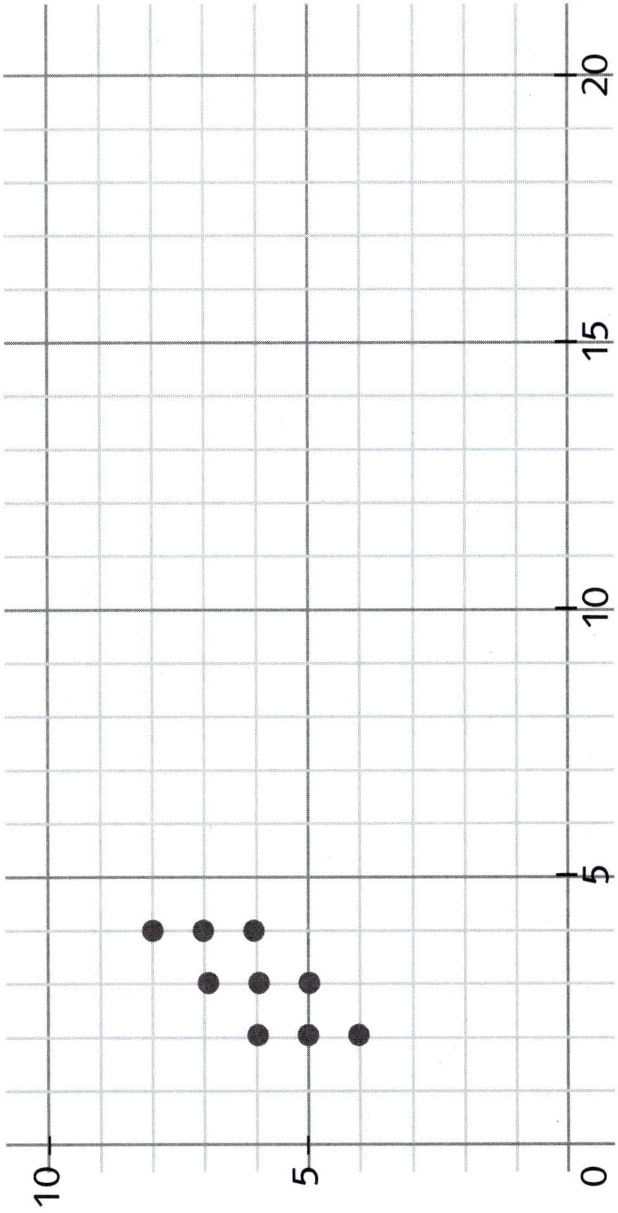

For use with Chapter 8, Question 27, section Interpreting Categorical and Quantitative Data.

REPRODUCIBLE 64

Scatterplots A–I

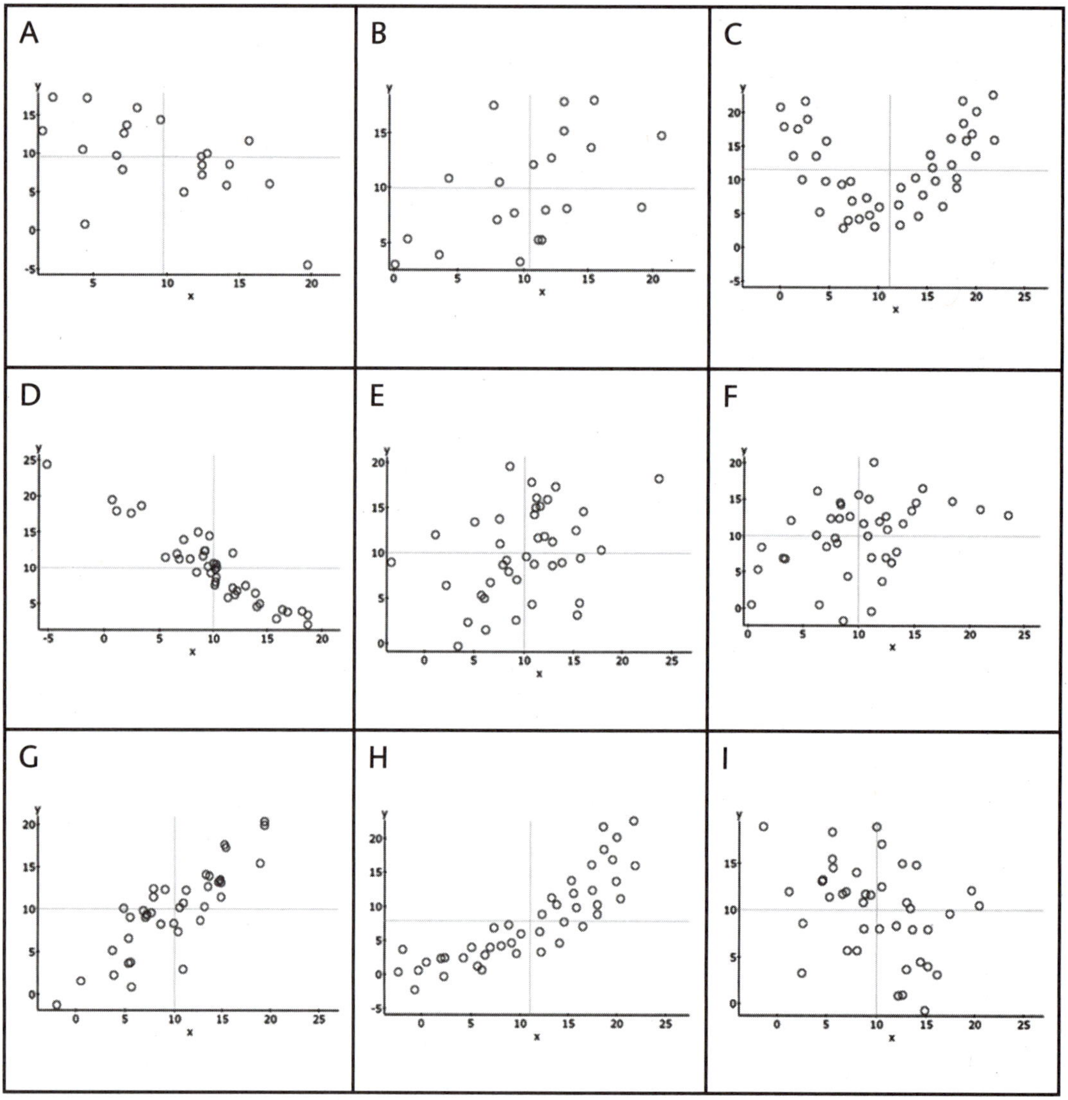

For use with Chapter 8, Question 29, section Interpreting Categorical and Quantitative Data.

Reproducible 65

7. Oologist's Dot Plots

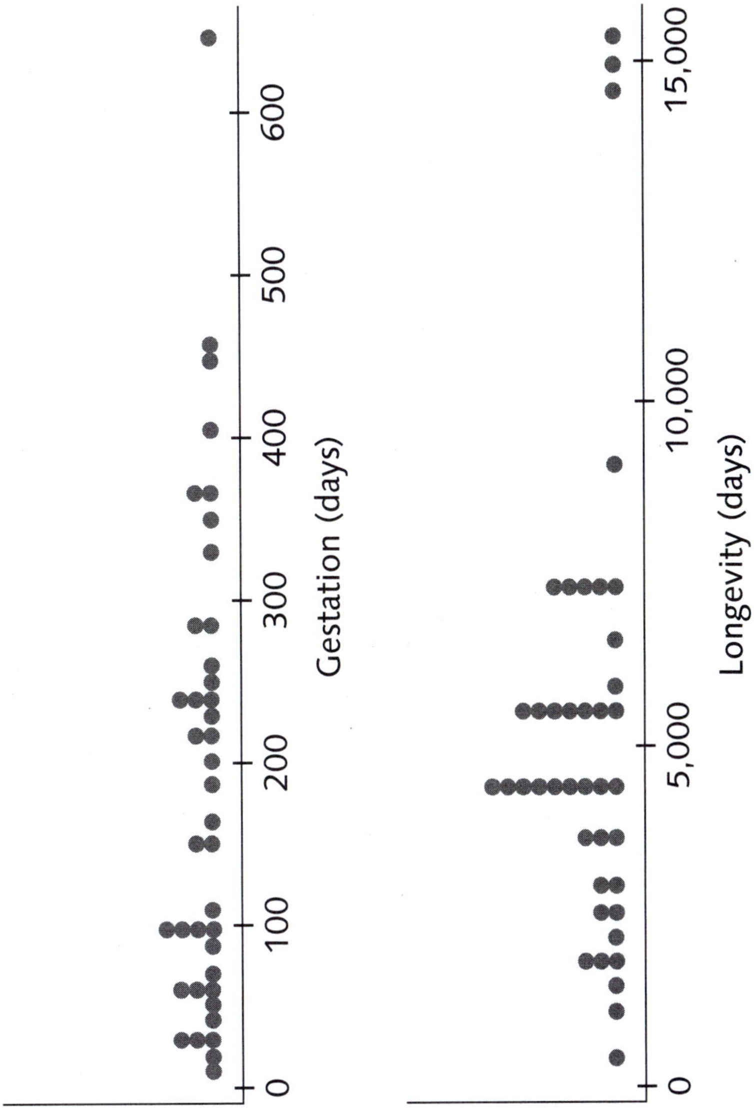

Data is based on shared data sets found on www.statcrunch.com.

For use with Questions 31–34, Chapter 8, section Interpreting Categorical and Quantitative Data.

Reproducible 66

7. Oologist's Scatterplot

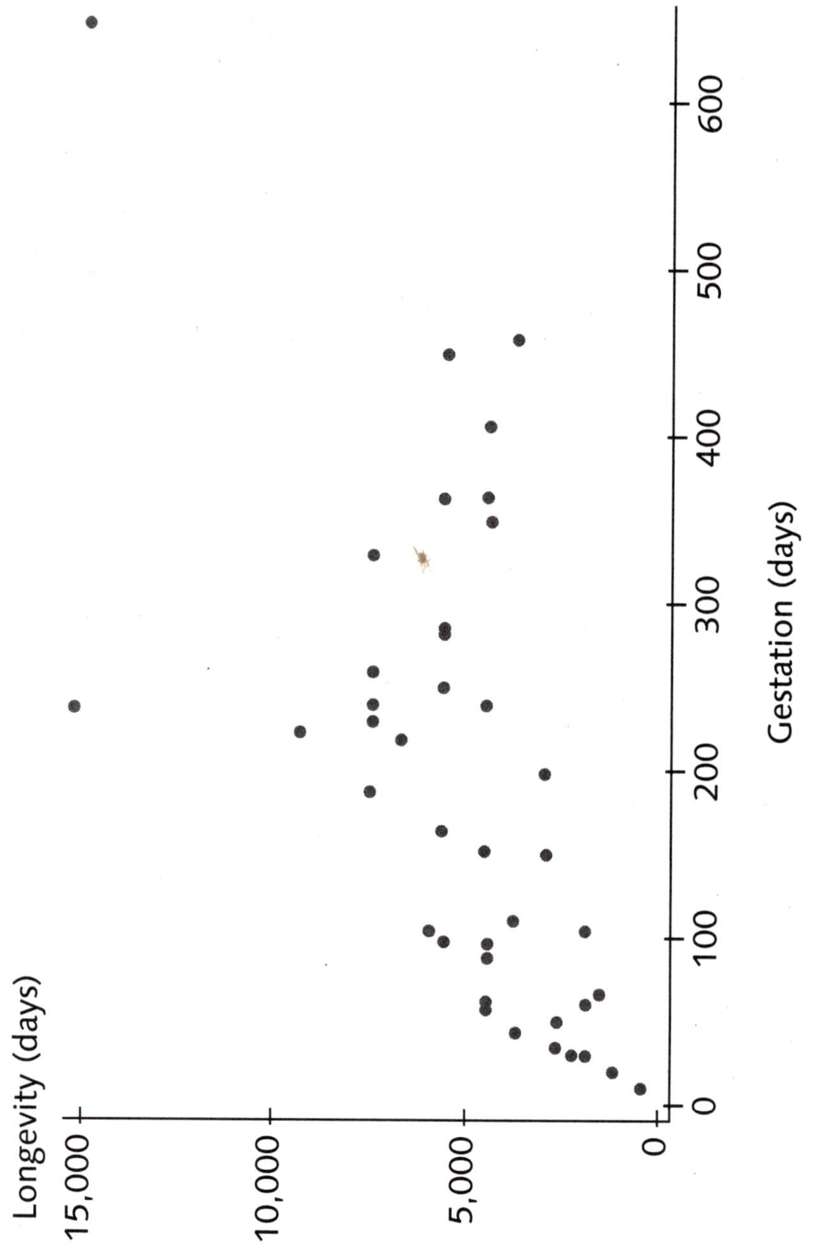

For use with Chapter 8, Questions 32 and 33, section Interpreting Categorical and Quantitative Data.

REPRODUCIBLE 67

Scatterplot and Linear Model

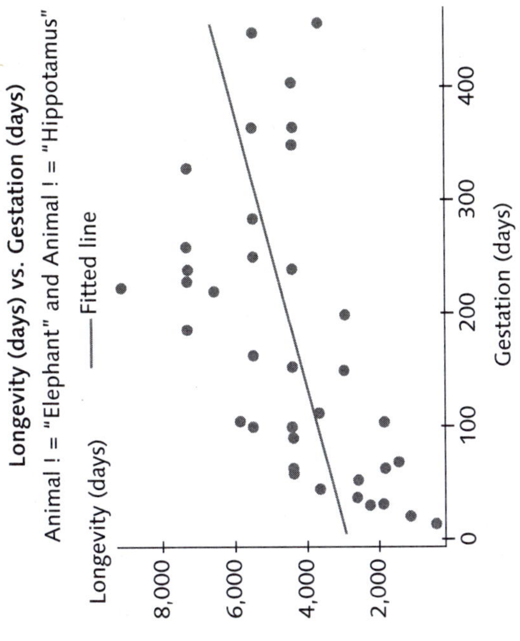

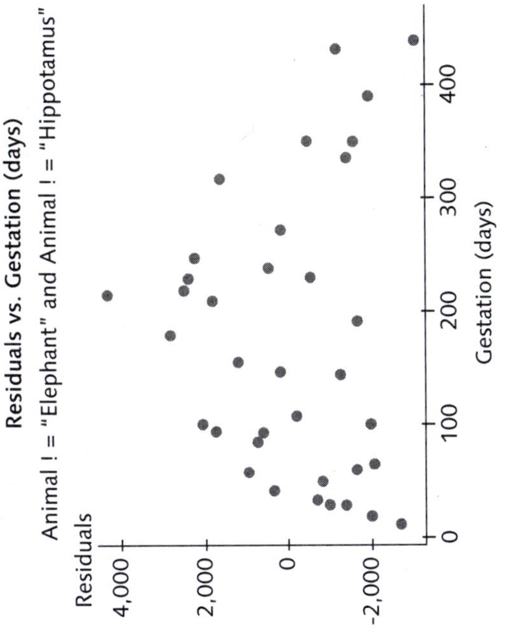

For use with Chapter 8, Question 34, section Interpreting Categorical and Quantitative Data.

REPRODUCIBLE 68

Scatterplot with Data

Explanatory	Response
2	5
6	9
6	6
7	8
7	3
12	12
13	10
14	13
15	19
18	15
$\bar{x} = 10$ $s_x = 5.077$	$\bar{y} = 10$ $s_y = 4.876$

Scatterplot

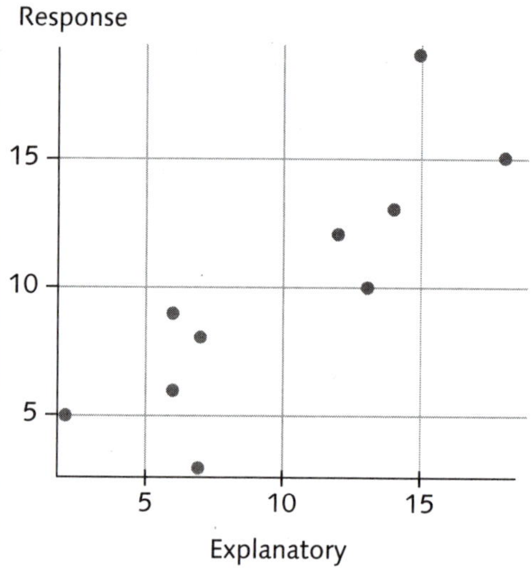

Correlation = 0.844

For use with Chapter 8, Question 36, section Interpreting Categorical and Quantitative Data.

Reproducible 69

Earnings of Domestic and Worldwide Movies

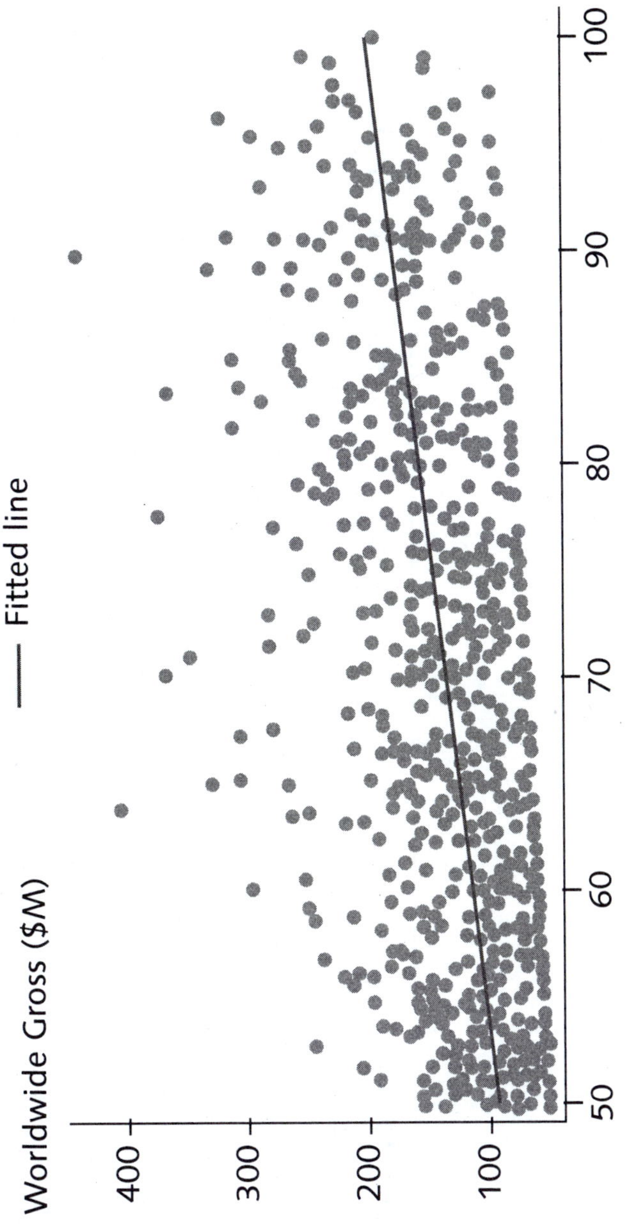

For use with Chapter 8, Question 38, section Interpreting Categorical and Quantitative Data.

Reproducible 70

Results of Card-Deck-Sorting Study

Sample	Mean	Standard Deviation
Sort Time (in seconds) Hearts and Diamonds versus Clubs and Spades	53.836667	3.3031853
Sort Time (in seconds) Hearts and Spades versus Clubs and Diamonds	62.123333	5.7704881

For use with Chapter 8, Question 7, section Making Inferences and Justifying Conclusions.

REPRODUCIBLE 71

Table of Qualities Important to Fourth- through Sixth-Grade Students

		Grade			
		4	**5**	**6**	**Total**
Goal	Grades	63	88	96	247
	Popular	31	55	55	141
	Sports	25	33	32	90
	Total	119	176	183	478

For use with Chapter 8, Question 7, section Conditional Probability and the Rules of Probability.

Reproducible 72

Probability Venn Diagram

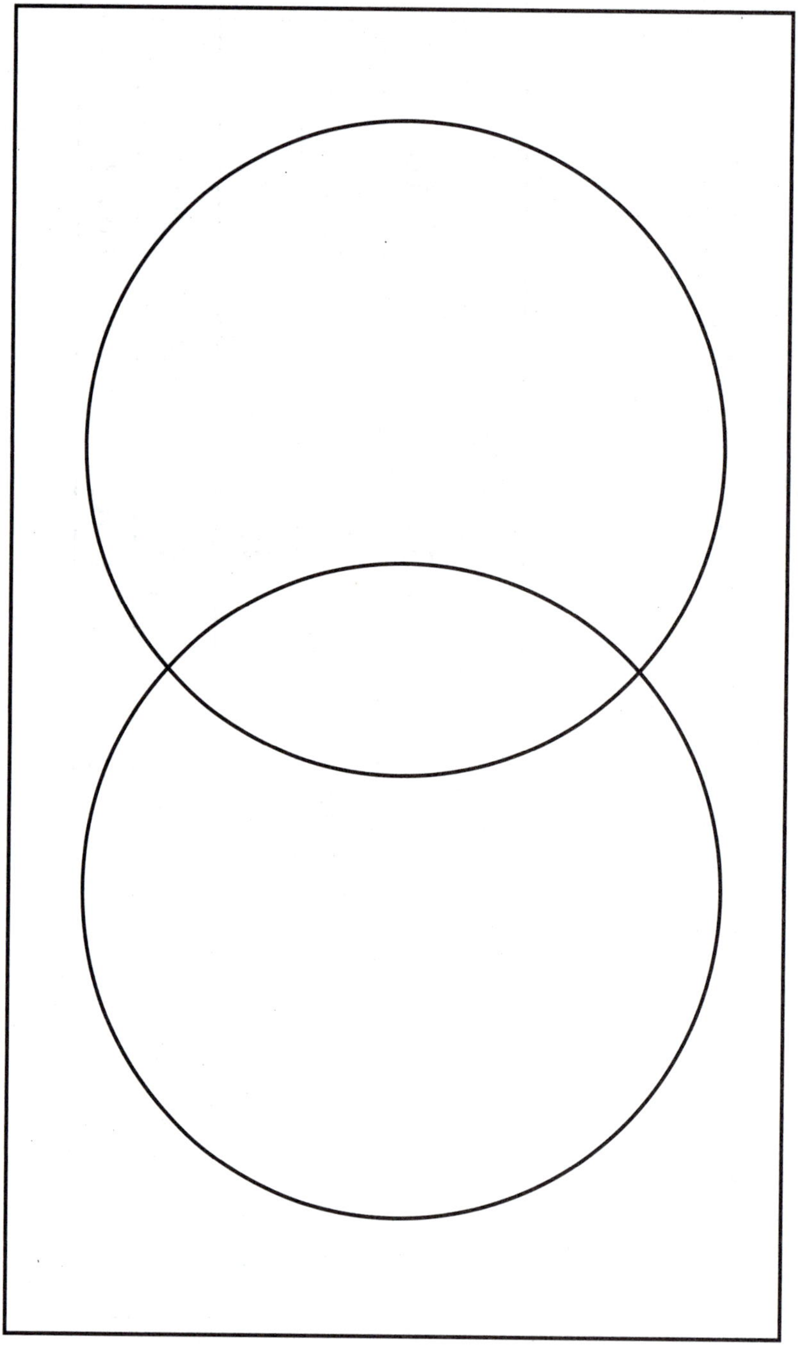

For use with Chapter 8, Question 9, section Conditional Probability and the Rules of Probability.

Reproducible 73

Kayla's Table

x	$f(x)$
0.1	0
0.01	0
0.001	0
0.0001	0
0.00001	0

For use with Chapter 9, Question 1, section Limits and Continuity.

Reproducible 74

Limit Statements from Graph

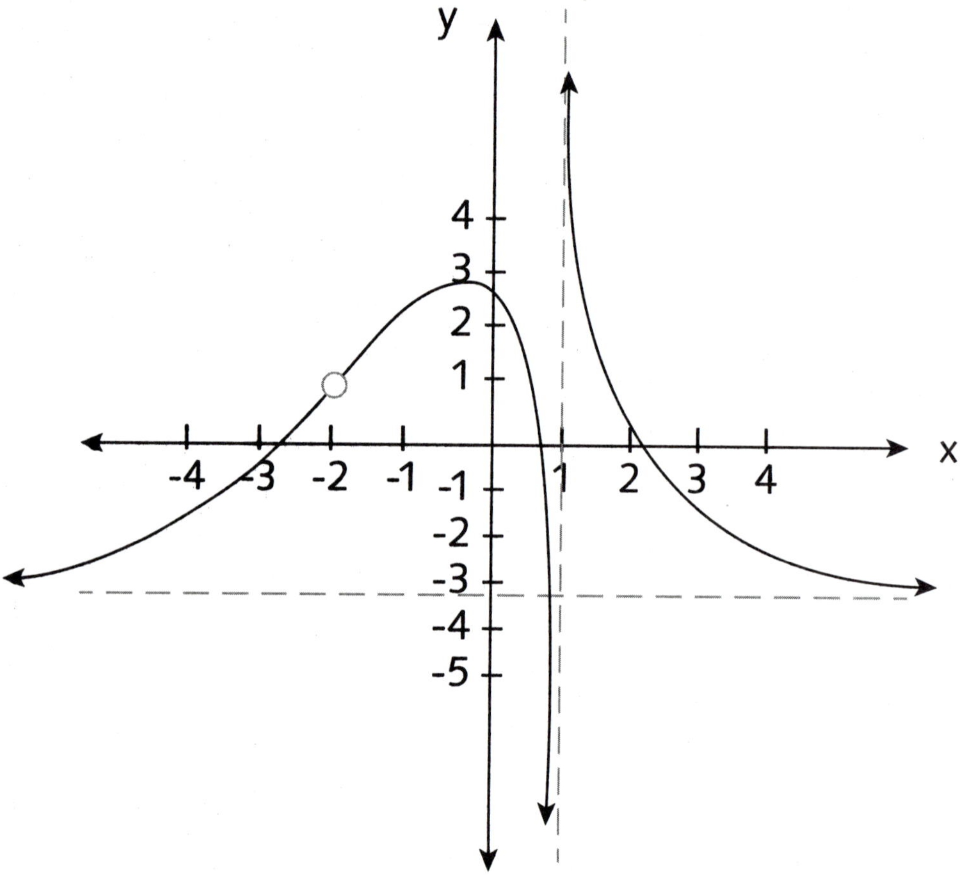

For use with Chapter 9, Question 6, section Limits and Continuity.

From *Good Questions for Math Teaching, High School* by Nancy Anderson, Leslie Dietiker, and Gregg Reilly.

Price of Ride-Share Service Graph

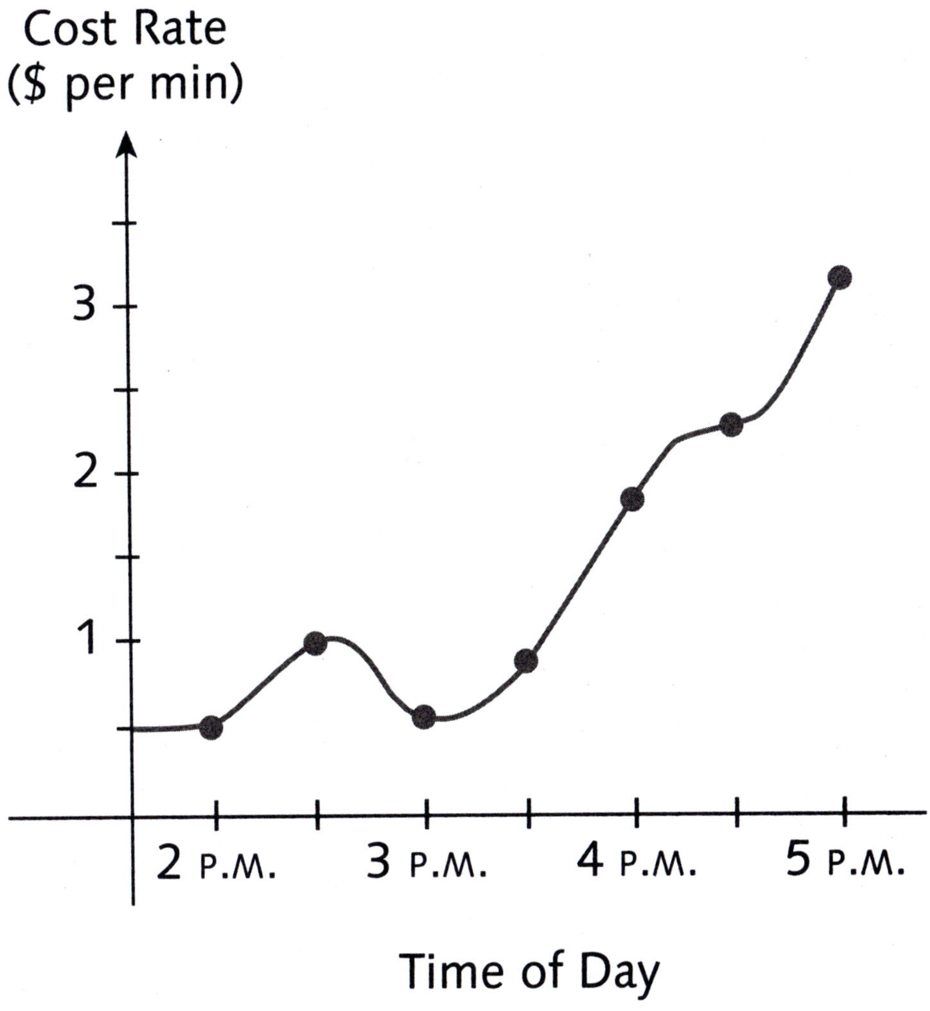

For use with Chapter 9, Questions 1–3, section Area Under the Curve.

Reproducible 76

Aria's Notes

- I had to double back at least once during my trip home.
- I was traveling faster at the beginning of my trip, but I got tired and slowed down.
- The trip took 30 minutes.
- I live 1 mile from school.

For use with Chapter 9, Question 3, section Slope of Tangents, Derivatives.

REPRODUCIBLE 77

Arthur's Graph Design

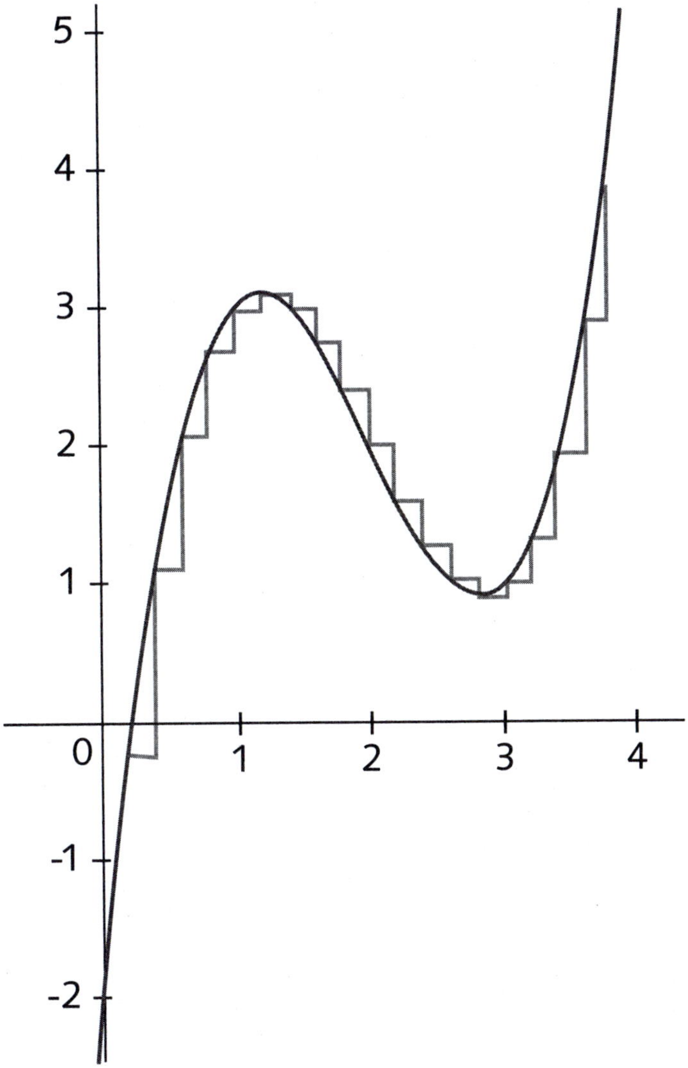

For use with Chapter 9, Question 4, section Slope of Tangents, Derivatives.

Reproducible 78

Roberto's Graph

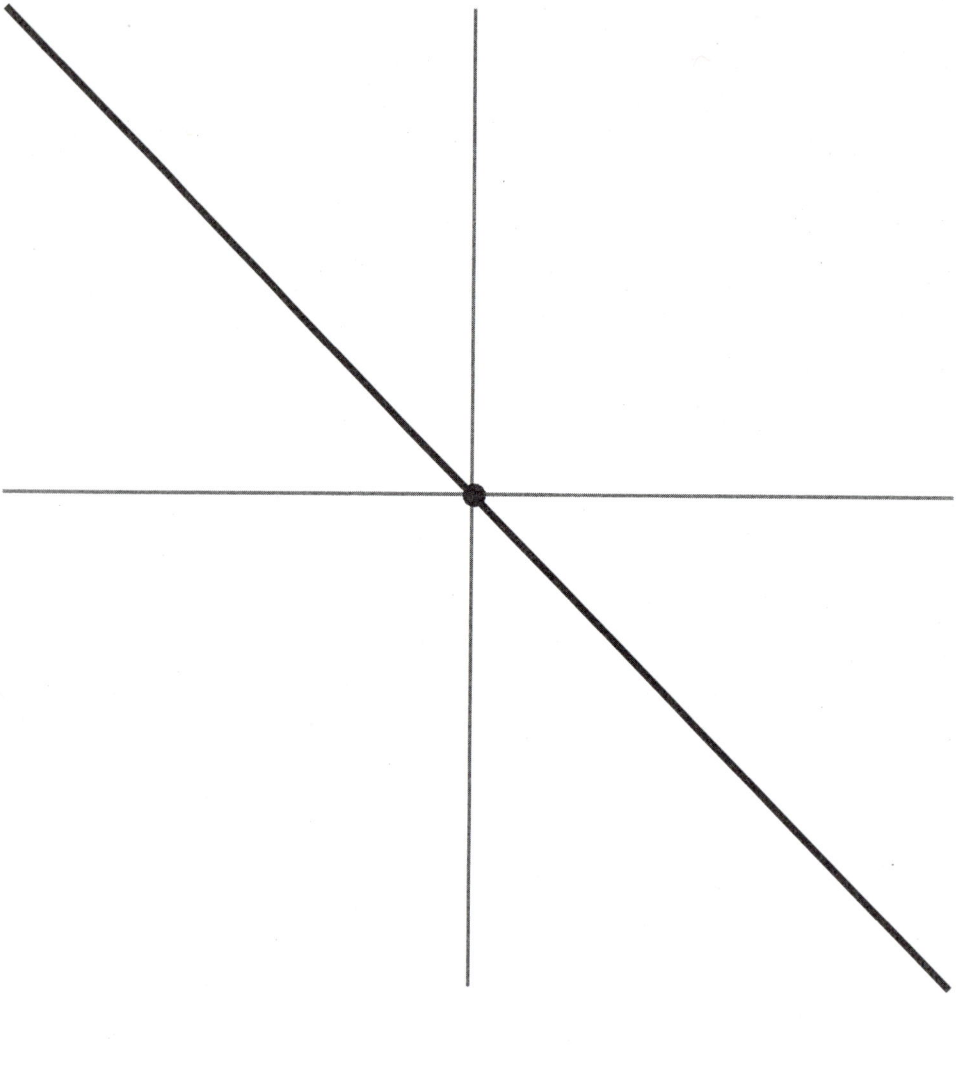

For use with Chapter 9, Question 4, section Slope of Tangents, Derivatives.

Reproducible 79

Venn Diagram with Curve Features

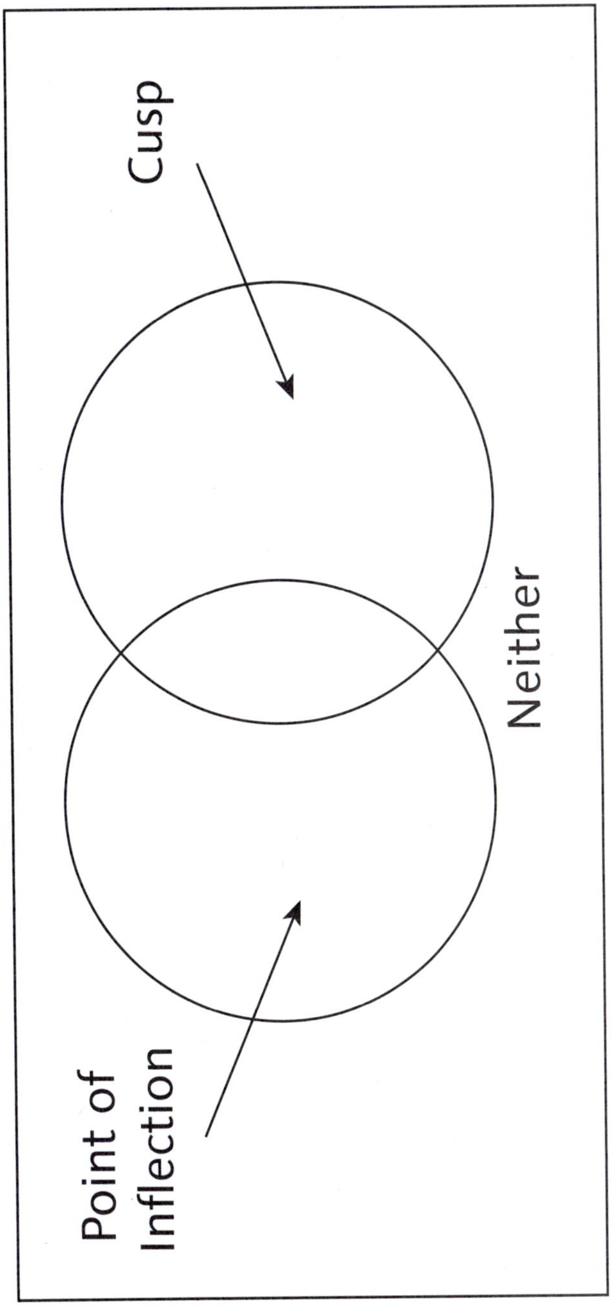

For use with Chapter 9, Question 6, section Slope of Tangents, Derivatives.

Reproducible 80

Keniesha's Graph

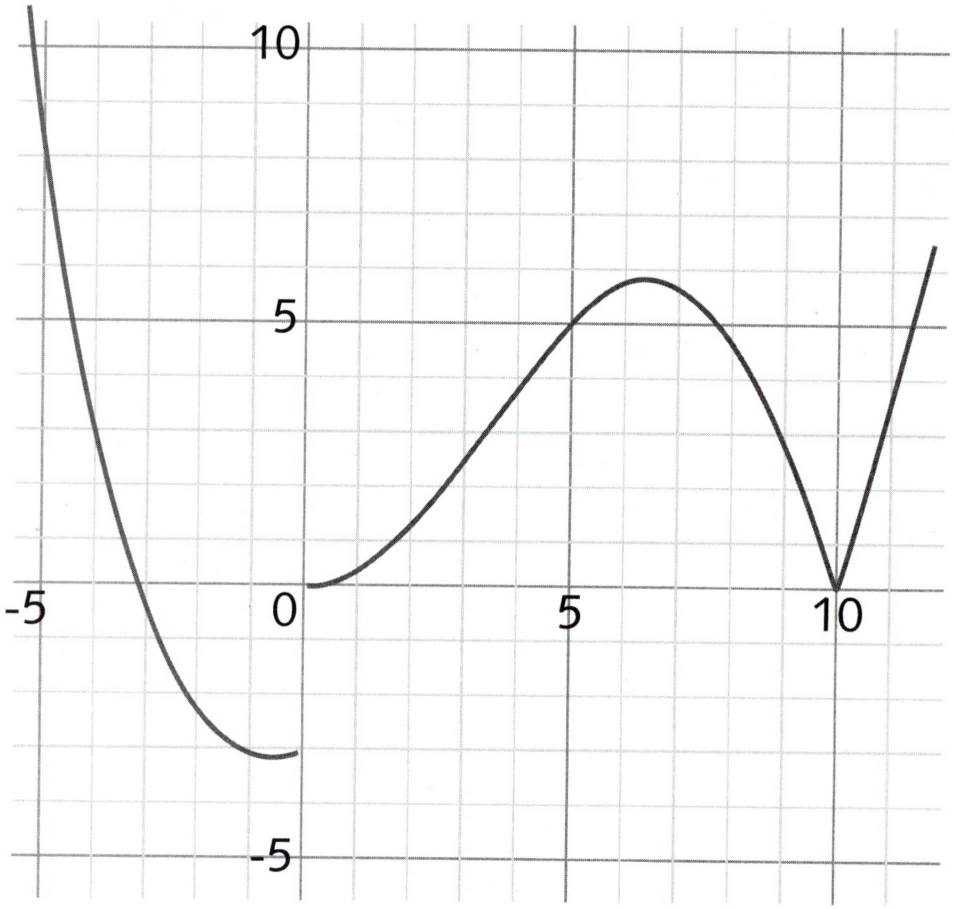

For use with Chapter 9, Question 8, section Slope of Tangents, Derivatives.

Reproducible 81

Motion Detector Graph

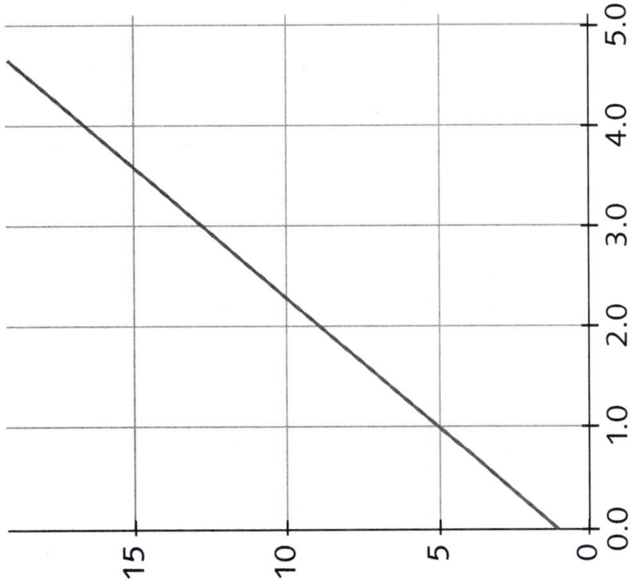

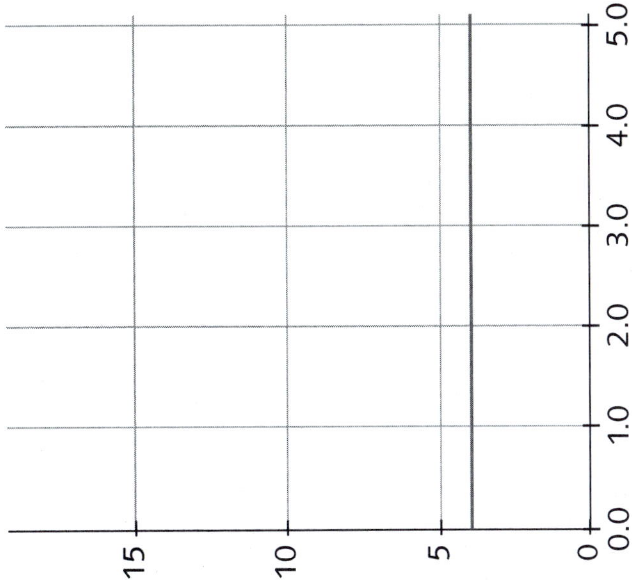

For use with Chapter 9, Question 3, section Fundamental Theorem of Calculus.

References

Brown, R. 1997. *Advanced Mathematics: Precalculus with Discrete Mathematics and Data Analysis*. Boston: McDougal Littell/Houghton Mifflin.

Chapin, S.H., C. O'Connor, and N. Anderson. 2013. *Talk Moves: A Teacher's Guide for Using Classroom Discussions in Math, Grades K–6*. 3d ed. Sausalito, CA: Math Solutions.

CME Project Development Team. 2013. *Algebra 2 Common Core*. Boston: Pearson.

Cooney, T.J., S. Beckmann, and G.M. Lloyd. 2010. *Developing Essential Understanding of Functions for Teaching Mathematics in Grades 9–12*. Essential Understanding Series. Reston, VA: National Council of Teachers of Mathematics.

Costa, A.L., and B. Kallick, eds. 2000. *Activating and Engaging Habits of Mind*. Alexandria, VA: Association for Supervision and Curriculum Development.

Dantonio, M., and P.C. Beisenherz. 2001. *Learning to Question, Questioning to Learn: Developing Effective Teacher Questioning Practices*. Needham Heights, MA: Allyn and Bacon.

DeVeaux, R., Velleman, P.F., and D.E. Bock. 2018. *Intro Stats*. 5th ed. Boston: Pearson.

Lloyd, G., B.A. Herbel-Eisenmann, and J.R. Star. 2011. *Developing Essential Understanding of Expressions, Equations, and Functions*. Essential Understanding Series. Reston, VA: National Council of Teachers of Mathematics.

National Council of Teachers of Mathematics. 2016. *Principles and Standards: Geometry*. Reston, VA: Author. https://www.nctm.org/Standards-and-Positions/Principles-and-Standards/Geometry/. Accessed June 9, 2020.

National Governors Association Center for Best Practices and the Council of State School Officers. 2010. *Common Core State Standards Initiative: Common Core State Standards for Mathematics*. www.corestandards.org/Math/Practice.

Schuster, L., and N. Anderson. 2020. *Good Questions for Math Teaching, Grades 5–8*. 2d ed. Boston: Math Solutions.

StatCrunch.com. "Animal Gestation and Longevity." Retrieved from: https://www.statcrunch.com/app/index.php?dataid=3470989. Accessed date: July 21, 2020.

USA Today. 2017. Major League Baseball Player Salaries. https://www.usatoday.com/sports/mlb/salaries/2017/player/all/. Accessed date: June 9, 2020.